A Biblical Tree Commentary

William Mitchell

Faithful Life Publishers
North Fort Myers, FL 33903

FaithfulLifePublishers.com

A Biblical Tree Commentary

Copyright © 2013   William Mitchell

ISBN: 978-1-937129-69-9

Published by:
Faithful Life Publishers
3335 Galaxy Way
North Fort Myers, FL   33903

www.FaithfulLifePublishers.com
info@FaithfulLifePublishers.com

Scripture quotations from the King James Version unless otherwise noted.

All rights reserved. No part of this publication may be reproduced, stored in a retrieval system, or transmitted in any form or by any means—electronic, mechanical. photocopy, recording, or any other—except for brief quotations in printed reviews, without the prior permission of Faithful Life Publishers and/or William Mitchell.

Printed in the United States of America

18   17   16   15   14   13   1 2 3 4 5

Table of Contents

Acknowledgement

First and foremost I thank the Lord Jesus Christ.

I would also like to thank Dr. Bryce Ausburger, our former pastor who encouraged me to write this book. There have been former pastors and men from my youth who have been sound Bible teachers that I have learned from, Dr. John R. Rice, Dr. Noel Smith, Dr. Rick Carter, Dr. Walt Handford, and of course my Dad who was also a pastor for many years.

In my professional career, there have been many who have instructed me in the dendrology field of sciences. Some of these have been Dr. Don VanOrmer, Dr. Alex Shigo, Dr. Reinee Hildebrandt, and Dr. George Ware.

Last of all, I would like to thank Pastor Huffman and his family, especially Mary, who have done the typing, editing, and compiling into a readable format.

Introduction

In this study, I have concentrated mostly on trees and woody plants, rather than all plants, for three reasons. <u>First</u> is my expertise in dendrology, as opposed to other branches of botany. <u>Second</u> is that in the Bible, the word for "tree" can also be translated "wood," and so woody plants fit in with the Biblical description of a tree. <u>Third</u> is that woody plants are more useful in that they can be used not only for the fruits that they naturally produce, but also as a wood product.

The use of the wood is illustrated well by the shittah, or acacia, tree, whose wood is very useful as a building material. The use of the fruit is illustrated well by such trees as the olive and fig.

As we progress through this study, it will become apparent that for some trees in the Bible, the wood is more important than the fruit. Good examples of such trees are the shittah, the cedar, and the cypress. But we will also see that for other trees, the fruit is more important than the wood. Good examples of these are the fig, the mulberry, and the sycamine. For still other trees, both are important. A good example of this kind is the olive tree.

But before we go any further, we must first stop and define what we mean by the word "tree." If this study is concentrating on trees, what criteria will we use to distinguish trees from other plants? The definition of the word "tree" varies very widely even among experts. In fact, there are about ten different definitions of a true tree.

The most narrow definition is a woody plant growing from a single stem usually to a height of over two meters (six feet, six and three-fourths inches). Under this definition, there are at least 21,000 trees.

On the other end of the spectrum, the broadest definition is a woody perennial plant having a single or multi stem with few branches on its lower part. Under this definition, there are as many as 100,000 trees. For this book, we will go with this wide definition since there are comparatively few trees in the Bible.

We will include a few non-woody plants in our study for various reasons.

As we begin this study, examining each of the trees of the Bible in detail, we will find that our interpretations many times differ from the opinions of Bible commentators. But this should not automatically cause alarm. Commentators are not botanists, and so due to their lack of expertise in this area, they sometimes take the names of trees out of context. Also, they often look at the growing conditions of the present barren land of Israel without considering the lush land *"flowing with milk and honey" (Exodus 3:8)* that it was in Old Testament times, before Israel's sin brought God's curse of barrenness and dryness upon the land.

On the other hand, most botanists have the opposite problem. They know a lot about plants, but have an evolutionary view, rather than a Biblical view.

We will try our best to lay aside both of these twin hindrances and examine each tree honestly, looking carefully at the Hebrew, Greek, and Latin words. With such close and honest investigation, we can better discover the true nature of the trees of the Bible.

In purely natural terms, the trees of the Bible make a particularly good and interesting study simply because of the land in which they grow, the land of Israel. There is no place on earth that has such diversity of climate in such a small area as this land has. Therefore, by studying the trees of the Bible, the trees that grow in this unique land, we are able to study a great variety of trees within relatively narrow borders.

But the real aim of this book is to go beyond physical discovery to spiritual application. For this reason, it concentrates on drawing applications from each tree, combining knowledge of its particular habitat, properties, and uses with an examination of its context in its verse, its chapter, and the Bible as a whole.

As we now enter this study, let us tune our hearts to be teachable by looking at a very fitting application drawn from wood. Wood is easier to work when it is green because its water and sap are still in it. Once it dries, it is much harder to work. In the same way, we as Christians are much easier to teach when our hearts are tender and sensitive to God's Word. When we allow our hearts to harden and our sensitivity to lessen, then we, like dry wood, become very hard to teach.

When this hardening happens to wood, it requires heat and water to bring it back to a condition in which it can be reshaped. When the hardening happens to us as Christians, it takes trials and tribulations to soften our hearts so that we can be reshaped into the Christians that Christ wants us to be. Let us spare ourselves this heartache. Rather than allowing our hearts to become hard, let us keep them tender and soft and sensitive to God's Spirit.

Especially as we begin this study, let us each one search our hearts and make sure they are in such a soft condition so that we can be easily taught and shaped by the Holy Spirit of God. Let us pray that, through this study, the Lord will teach us many important spiritual truths from the trees of the Bible, and that thereby our lives may be fashioned into all that He desires them to be.

1
The Algum Tree and the Almug Tree

We begin our study of the trees of the Bible with two trees that I have been pondering for some years, the algum and the almug. Many commentaries say that they are the same tree, but I believe there is a distinction between the two when they are carefully examined.

The algum tree is mentioned in II Chronicles 2:8. *"Send me also cedar trees, fir trees, and algum trees, out of Lebanon: for I know that thy servants can skill to cut timber in Lebanon; and, behold, my servants shall be with thy servants." (II Chronicles 2:8)* From this verse, it is clear that the algum tree came out of Lebanon.

When it is mentioned later on, it seems at first that it came from Ophir. *"And the servants also of Huram, and the servants of Solomon, which brought gold from Ophir, brought algum trees and precious stones. And the king made of the algum trees terraces to the house of the LORD, and to the king's palace, and harps and psalteries for singers: and there were none such seen before in the land of Judah." (II Chronicles 9:10-11)* But a careful look reveals that only the gold is said to come from Ophir. Those who brought the gold also brought the algum trees, but the verse does not say that the algum trees themselves came from Ophir.

In I Kings, the al**mug** tree is mentioned, and it does come from Ophir. *"And the navy also of Hiram, that brought gold from Ophir, brought in from Ophir great plenty of almug trees, and precious stones. And the king made of the almug trees pillars for the house of the LORD, and for the king's house, harps also and psalteries for singers: there came no such almug trees, nor were seen unto this day." (I Kings 10:11-12)* The reading of this passage is markedly similar to that of II Chronicles. Both accounts record the coming of the Queen of Sheba, the construction of the temple, and the bringing in of

9

spices. For this reason, most commentaries say that they are the same tree.

In order to hold to this claim, some commentators say that the Bible translators transposed the letters "g" and "m," "gimel" and "mem" in Hebrew. But this leads to a dangerous position. If they switched these letters, what else in the Bible might they have switched?

We know, of course, that God often changed people's names in the Bible. He changed Abram to Abraham and Saul to Paul. Also names of cities were changed. Jacob changed the name of the city of Luz to Bethel after seeing the vision of the ladder and angels. Other cities had their names changed as a result of military conquest.

But the names of animals and plants are unchanged in the Bible. The reason for this is that God works with people, not with animals and plants. When God changes a person, He often changes that person's name. When the devil changes a person, he also changes the name. Satan sought to change Daniel into a pagan by having Nebuchadnezzar change his name from Daniel, "God is my judge," to Belteshazzar, "Bel will judge," using the name of a pagan god. Cities, which are groups of people, are also changed as God works with the people, and their names are changed accordingly. But the names of plants and animals remain unchanged.

A closer look at the two Biblical accounts reveals that, while they are very similar, they are not exactly the same. II Chronicles makes the temple prominent and thus records the construction from God's point view, while I Kings makes the palace prominent and thus records it from man's point of view. These accounts are similar, but in no way identical. Could it be the same with the algum and almug trees? Perhaps they are very similar trees, but not identical.

Even today we can find examples of trees that have very similar or even identical names, but are different trees. The swamp oak and the swamp white oak sound similar, but they are distinct and different. The yellow chestnut is different from the chestnut. The yellow chestnut is also different from the yellow poplar. The yellow poplar in turn is different from the poplar.

On the other hand, there are cases in which the same tree can have different names. The terms "black gum" and "sour gum" are two different names for the same tree. Also the terms "Scotch pine" and "Scots pine" refer to the same tree.

But the Scots pine of northern Canada, though its name is identical with that of the Scots pine of North America, is a different tree entirely. The bluebell of Scotland is a bulb plant, but in England, the term "bluebell" refers to a completely different plant. The tree called a Georgia pine in Georgia is referred to as a southern pine in other states.

Because of the confusion that can quickly result from such an unorganized system, an official system of nomenclature was established in the 1700s. A unique Latin name was assigned to each tree. In this way, while the common names are still widely used among most people, confusion is eliminated among professionals by having a unique name for each kind of tree. Whereas it used to take up to twenty or thirty words just to describe one plant, the same plant can now be described with just one Latin name, allowing greater precision and ease of communication.

But we do not have the Latin name to lean upon to determine the nature of the algum and almug trees, which were so designated long before the new nomenclature. These English words are not a true translation, but a transliteration of the Hebrew. And even since the time of the transliteration, the spelling of "algum" has changed from "algume" to "algum." So in trying to discover exactly what type of trees they are in today's language, we are left with only speculation. Yet we can learn something of their properties by examining the ways in which they were used.

1. First, they would have had to have been precious wood. Otherwise, why spend the time and money to have them shipped in from far away?

2. They also would have had to have been strong, suitable to be used for terraces and pillars.

Most scholars and rabbis believe that the almug is red sandalwood. At first, this seems impossible because the red sandalwood of today is not hard enough to be used in such construction as pillars. But over time, plants do change. This change can be seen in growth rings. The slower a plant grows, the smaller its growth rings are, and the harder its wood is. But the faster it grows, the wider the growth rings, and the more flexible the wood. The growth rings on trees today are much wider than they were just a few hundred years ago, which means that the wood is more flexible today.

A few years ago, construction was begun to build a replica of a historical ship. The original ship had had a sixty-foot-tall mast made of yellow pine. In order to be authentic, the men who were working on the ship began searching for a yellow pine that would be tall enough, hard enough, and flexible enough to serve as a mast. They found plenty of yellow pines, but had great difficulty finding one that fit this description. The fast growth of recent years had made it so that the wood of this pine was softer than in years past when the original ship had been built. Though new yellow pines were all around, the men were forced to search all over for an old pine tree.

But these examples show the change over just a few hundred years. Given the at least two thousand years of change since Biblical days, it is certainly likely that the red sandalwood would have been much harder then than it is today. It could no doubt have been hard enough to be used for pillars.

3. Another property that both the algum and almug trees would have had is beauty. They were both used for making musical instruments, harps and psalteries. For such high quality instruments as these, we use beautiful wood, perhaps with a red grain running through it, and strong wood that allows it to resonate with a beautiful, rich, and full tone. These trees would have had a very desirable and valuable wood. Only then would they have been worth importing for such uses.

If the almug is possibly red sandalwood, then what is the algum? Of course, some say that both are red sandalwood. Others, claiming that both are similar but not identical, claim that while the almug is red sandalwood, the algum is white sandalwood.

Another problem that we have in studying these trees is that no one knows exactly where Ophir was. India, West Africa, and

Saudi Arabia have all been offered as possibilities. But the lack of concrete knowledge about this place and its botanical environment leaves us largely to speculation.

To conclude, it is my personal opinion that the algum and almug are two different trees. They come from two different places, Lebanon and Ophir, and they have different spellings. The Bible is very precise and careful with detail, even down to spelling. Every letter is there for a reason. Just because two words sound and look very similar does not mean that they are the same. This same principle can be seen in two other difficult trees, the sycamine and sycamore, which we will look at later.

In fact, the very next tree we will look at, the almond tree, has a name that sounds very similar to the name "almug tree." Yet no one suggests that those two are the same. That is because we are familiar with the almond tree and its properties. But might not the same be true of trees with which we are not familiar as well? Let us no longer be confused by these trees, but take them as the Bible seems to give them, as separate and distinct trees.

2
The Almond Tree

The almond tree appears in the book of Ecclesiastes in a passage describing old age. *"Also when they shall be afraid of that which is high, and fears shall be in the way, and the almond tree shall flourish, and the grasshopper shall be a burden, and desire shall fail: because man goeth to his long home, and the mourners go about the streets." (Ecclesiastes 12:5)*

The almond tree is very fitting for this passage on old age, because in many ways it is the symbol of old age. It bears white flowers very early in the springtime. In full bloom, the almond tree looks like the hoary head of an aged person, reminding us all of old age, death, and eternity.

Its early blooming is another part of this symbol. Because it is one of the first trees to bloom, it announces to the world that spring is coming and that other plants will quickly bloom. For this reason, it is often used in the Bible as the herald of things that are soon to come. In this context of old age, its white flowers symbolize white hair and signal that eternity is soon to come.

As every older person has found out, time seems to move faster the older we become. This is based upon our time reference. At three years old, one year is a whole third of our lives and seems very long. But at thirty, it is only a thirtieth and doesn't seem so long anymore. By the time we are seventy, it is a seventieth of our lives and seems to go by with a flash. Thus for the elderly, future events seem to approach more quickly than they used to, and death and eternity are very near. The almond tree with its head of white flowers pictures both this old age and the quick approach of things to come.

In Jeremiah, the almond is used as just such a herald. *"Moreover the word of the LORD came unto me, saying, Jeremiah, what seest thou? And I said, I see a rod of an almond tree." (Jeremiah 1:11)* Here the almond tree announces the speedy judgment coming upon Israel. *"Then said the LORD unto me, Thou hast well seen: for I will hasten my word to perform it." (Jeremiah 1:12)*

The image of the almond was also to be used in the tabernacle. The candlestick was to have *"three bowls made like unto almonds, with a knop and a flower in one branch; and three bowls made like almonds in the other branch, with a knop and a flower: so in the six branches that come out of the candlestick. And in the candlestick shall be four bowls made like unto almonds, with their knops and their flowers." (Exodus 25:33-34)*

When the candlestick was fashioned, it was made exactly according to this divine prescription, with the bowls made like almonds. *"And he made the candlestick of pure gold . . . Three bowls made after the fashion of almonds in one branch, a knop and a flower; and three bowls made like almonds in another branch, a knop and a flower: so throughout the six branches going out of the candlestick. And in the candlestick were four bowls made like almonds, his knops, and his flowers." (Exodus 37:17-20)*

This candlestick and its almond-shaped bowls were a picture of Jesus Christ. As their function was to give light, so He said of Himself *"I am the light of the world." (John 8:12)* Also they were made of solid gold. They were not wood overlaid with gold, as were some of the items of the tabernacle, but pure gold. This also pictures Christ, for just as they were pure and without impurities, so Jesus Christ is perfect and free from all impurities.

Another place where the almond appears is in Numbers 17. Here the people had just witnessed the rebellion of Korah and his horrifying judgment. But even after this, they had murmured against the authority of Moses and Aaron. To silence this murmuring, the Lord had commanded that a contest be held. The prince of each tribe brought a rod to the tabernacle. Aaron being the prince of the tribe of Levi, his rod was brought for that tribe. The Lord would then declare His choice of an authority by causing one of these dead rods to bud, the rod of the man of His choosing. In verse 8, we read of God's choice. *"And it came to pass, that on the morrow Moses went into the tabernacle of witness; and, behold, the rod of Aaron for the house of Levi was budded, and brought forth buds, and bloomed blossoms, and yielded almonds." (Numbers 17:8)*

To understand this magnitude of this miracle, we must understand something of the nature of the almond tree. Not only was this rod a dead rod, but the almond tree is not self fertilizing. It needs two separate trees in order to fertilize and produce fruit. Yet Aaron's rod budded and brought forth fruit without any other live almond tree around. In this way, it was a double miracle and a clear and undeniable act of God. Another thing to consider is that this rod may not have even been from an almond tree. If this were the case, then it was a triple miracle. Clearly, in any case, this was a miracle that could be accomplished only by the supernatural power of God.

In this miracle, we see another picture of Christ's ability to restore Israel. The rod was only a dead piece of wood, but God was able to give life to it. Indeed He is the only One Who can give life to a dead piece of wood, for He alone holds the power of life and death. Just as He cursed the fig tree in Mark 11, making a green tree dry, so He can also do the opposite and make a lifeless tree flourish again. Though the Jewish nation today seems dead like an old piece of wood, yet one day Christ will give it new life, and it will bloom again through the power of Christ.

But the greatest symbol of the almond is its symbol of Christ Himself. With its white flowers and its being the symbol of age, it represents Christ as He appears in Daniel 7, as the *"Ancient of days." (Daniel 7:9, 13, 22)* Christ Who has no beginning or end has existed since eternity past. He has lived through every age and continues to this day. In Revelation, this same picture of age is maintained as Christ's hair is described as being white. *"His head and his hairs were white like wool, as white as snow." (Revelation 1:14)* While Christ is ever fresh and full of new life, like the new white blossoms of the almond, He is also ancient and of great age.

The almond also, as the herald of things soon to come, proclaims throughout the world that Christ's second coming is quickly approaching. From our perspective, it seems that we have been waiting long for Christ's return, as we watch the world continue to get worse. In fact, we often hear the phrase "Christ is tarrying His coming." But in reality, Christ is not tarrying. Hebrews says plainly, *"For yet a little while, and he that shall come will come, and will not tarry." (Hebrews 10:37)* Christ has an appointed time at which He will come, and it is not far in the future, but only *"a little while."* It only seems long from our point of view. When Christ comes, He will come quickly, just as when we see the almond flowers, we know spring is at hand.

For us as Christians, this soon return of our Lord should be a prick in our side to motivate us to action. Just as the white flowers of the almond tree proclaim Christ's coming, they also remind us of our own responsibility in light of that coming. With their whiteness, they bring to our minds the words of Christ, *"Lift up your eyes, and look on the fields; for they are white already to harvest." (John 4:35)* There is a great harvest ready to be reaped, but the time is short. We are very close to the coming of Christ. His time is at hand, and therefore we have very little time in which to reap this harvest. Let us hearken to the call of the almond tree and make use of the time we have remaining by reaping this harvest that is white with ripeness.

3
The Aloe Plant

The aloe plant is not mentioned a lot in Scripture, but it is mentioned a few times. The first mention is in Numbers, when Balak calls Balaam to curse the Israelites Though Balak wants Israel to be cursed and though Balaam wants to curse Israel, yet the Lord will not allow it. He causes Balaam to speak words of blessing against his own will.

Balaam says, *"How goodly are thy tents, O Jacob, and thy tabernacles, O Israel! As the valleys are they spread forth, as gardens by the river's side, as the trees of lign aloes which the LORD hath planted, and as cedar trees beside the waters."* (Numbers 24:5-6)

The word "lign" is today spelled lignin. It refers to the sap of any tree, which when combined with cellulose, makes up the woody tissue of that tree. It is this substance that gives wood its weight and hardness. An abundance of it makes a very dense and heavy tree, and a lack of it makes a light and porous tree. Thus lign is the essential part of the tree. It is the tree's sweetest, inner part that makes up its very substance.

As the word is used in this verse, its describes an aloe plant of great richness and density, and full of sweet sap. It describes it as heavy with sweet fragrance. It is to this sweet-smelling aloe plant that Balaam likens the tents of Israel. Though he desires to curse Israel, the Lord turns his words into a blessing. Against his intention, Israel is described as being sweet-smelling in the nostrils of Christ, just as sweet and fragrant as a rich aloe plant.

Besides its fragrance, the aloe plant carries in its leaf many other benefits to man. Perhaps the simplest and best known is its use in skin care simply through rubbing it on. We can today do this with aloe soap and aloe lotion, and thereby improve our skin. But also drinking aloe helps in respiration and digestion, and purifies the blood.

All these benefits are rooted in the center of the plant. The center of the aloe plant is its most important part. It is the part that is rich is lign, and the only part that is fragrant. It is what sustains the rest of the plant and gives it its properties. Without this vital center, the rest of the plant could not offer its life-giving benefits.

But this center takes hundreds of years to grow and develop. It is very costly and time-consuming to grow and cultivate a good aloe plant. For this reason, the lign, the essential center substance, of the aloe plant is very expensive. At one time, it was worth its own weight in gold.

The next Biblical reference speaks of ground aloe wood being rubbed into clothes. *"All thy garments smell of myrrh, and aloes, and cassia, out of the ivory palaces, whereby they have made thee glad." (Psalm 45:8)* In clothes, the aloes would have served two purposes. First they would have kept fleas and lice away, just as they do when rubbed on skin. And they would also have served as a perfume, giving a sweet fragrance to the garment and to the one who wore it.

Another place on which aloes were rubbed in Bible times was the bed. *"I have perfumed my bed with myrrh, aloes, and cinnamon." (Proverbs 7:17)* In those days, a bed was not made of a mattress like we have today. Instead, it was nothing more than a pile of old pieces of cloth. Just as aloes were rubbed onto clothes, they were rubbed onto these cloths as well. Here the aloes again served both as a perfume and as a means of keeping away fleas, ticks, and other insects. In this way, the people in Bible times kept themselves healthy and free from the diseases carried by insects.

With its all-important and precious center, the aloe plant illustrates well the principle that the heart of man is more important than his outward appearance. The aloe plant, in shape somewhat similar to a palm tree, yet lacking the compound leaves of the palm and having thicker and wider leaves, is not remarkable on the outside. While its leaves do play an important medicinal purpose, it is its heart that is most important.

While most of the references to the aloe plant are in the Old Testament, there is one reference to it in the New Testament. It is found at the account of Nicodemus' preparation for Jesus' burial. *"And there came also Nicodemus, which at the first came to Jesus by night, and brought a mixture of myrrh and aloes, about an hundred pound weight. Then took they the body of Jesus, and wound it in linen clothes with the spices, as the manner of the Jews is to bury." (John 19:39-40)*

Here another benefit of the aloe plant is found. Aloes here were used, not necessarily to keep bugs away, but for preservation. Mixed with such a substance as myrrh, the aloe plant can indeed preserve a dead body.

This mixture of myrrh and aloes that Nicodemus brought was said to be *"about an hundred pound weight."* The pound of Bible days was a little different from our pound today. Instead of being 16 ounces, it was about 12 ounces. Still, a hundred of even these slightly smaller pounds would have meant a very great amount of these spices. With the precious aloes already being expensive even in small amounts, this large amount would have been very, very costly.

We don't know exactly how Nicodemus obtained these spices. Perhaps he was rich enough himself to buy them. Or perhaps both he and the rich Joseph of Arimathaea, along with others of Christ's disciples, pooled their money together, and Nicodemus bought the spices with this pool. Of course, had these sincere men understood and believed the Scriptures and Christ's Own teaching that He would rise after three days, they could have saved this expense. They would have known that Jesus' body needed no preservation, for it would rise in three days.

But nevertheless, this great expense shows the importance and worth that these men placed on Christ. It reveals their profound love and respect for Him. It is symbolic also, not only of His value to these men at this time, but of His eternal and supreme value above all else in the world.

Indeed Christ is of priceless worth. He is worth our greatest expense. He is worth us dedicating our very lives to Him. As we do, our hearts will grow into something of precious worth and sweet fragrance in His sight as well.

4

The Apple Tree

The first thing that comes to many people's minds when they think of the apple tree in the Bible is the fruit that Adam and Eve ate. Popular tradition has it that this fruit was an apple. But this is highly improbable. The apple is mentioned by name several times in the Bible. If it were the forbidden fruit, then why would the Bible not name it there? The Bible does not say what kind of fruit it was that Adam and Eve ate, and therefore no one knows what it was.

There are plenty of clear references to apples in the Bible. We don't need to make the forbidden fruit into something the Bible does not claim for it to be. So we turn our attention to the places in the Bible where the apple is specifically named.

The Hebrew word for apple is *tappuah* (תַּפּוּחַ). In its most literal sense drawn from its root, it means sweet-scented. There are several places in the Bible where this word is simply transliterated into English.

Joshua 15 is one of these. This chapter names the conquered cities of the promised land that went to the tribe of Judah. Verse 53 continues the list with these cities, *"And Janum, and Bethtappuah, and Aphekah. (Joshua 15:53)* Bethtappuah, the name of the middle city, is a combination of two Hebrew words, *beth* which means "house," and *tappuah* which means "apple." Thus the name of this city is "House of Apples."

Joshua 16 continues naming cities conquered in the land of Canaan. This chapter names the cities that went to the tribe of Ephraim. It says that the border of this tribe, *"went out from Tappuah westward unto the river Kanah; and the goings out thereof were at the sea." (Joshua 16:8)* Again a town carries the name "Tappuah," "Apple." This same town is mentioned also in Joshua 12 where the conquered kings of the various Canaanite towns are named. *"The king of Tappuah, one." (Joshua 12:17)*

Joshua 17 continues naming cities, and names those of the tribe of Manasseh. *"And the border went along on the right hand unto the inhabitants of Entappuah." (Joshua 17:7)* This town begins with the word "en," which means "fountain." Thus its name in full means "Fountain of Apples." No doubt all of these towns with the name "tappuah" abounded with apple trees and apples.

Most Bible scholars say that the apple tree of the Bible is not the same as the apple tree we know today. Our apple trees are in the genus *Malus*. But these Bible scholars say that the Biblical apple trees were in the genus *Prunus*. They say that they were probably apricot trees or quince apple trees. But the references to the apple tree in Song of Solomon don't hold up to this interpretation. They don't seem to fit the apricot or the quince apple.

Chapter 2 of Song of Solomon says, *"As the apple tree among the trees of the wood, so is my beloved among the sons. I sat down under his shadow with great delight, and his fruit was sweet to my taste." (Song of Solomon 2:3)* This verse speaks of the apple tree as being a shade tree and having sweet fruit.

First, neither the apricot tree nor the quince apple tree is big enough to serve as a shade tree. Later in the book, this image is repeated where it says, *"I raised thee up under the apple tree." (Song of Solomon 8:5)* This tree must have been a broad tree, providing shade and a cool place to sit.

Second, Song of Solomon presents this tree as bearing sweet fruit, *"his fruit was sweet to my taste." (Song of Solomon 2:3)* Later it says, *"Stay me with flagons, comfort me with apples: for I am sick of love." (Song of Solomon 2:5)* This comfort also implies a sweet fruit and one with a sweet fragrance of perfume, comforting to the sick. This sweetness is certainly not in keeping with the quince apple. The quince apple is a very sour fruit, not sweet. But the Biblical apple was apparently sweet.

Another problem with the apricot being the Biblical apple is timing. For thousands of years, the apricot was cultivated only in China. It was not brought to Israel until around the 300s B.C., when Alexander the Great brought it. This was more than a thousand years after Joshua had conquered the cities of Bethtappuah, Tappuah, and Entappuah, known for their abundance of apples. It was also about seven hundred years after Solomon wrote his song in which he described the apple tree.

The quince apple would probably be closer than the apricot. But the strongest argument against it is its bitter and sour fruit. If it were the apple referred to in the Bible, then many Biblical references would be rendered inapplicable.

Another interesting fact about the quince apple is that it was sacred to the Greeks who worshiped the goddess Aphrodite. Her statue pictured her holding a quince apple. So the people of Bible times knew about the quince apple, but it was a fruit of pagan associations, and not of Biblical imagery.

Still an objection that is often raised to the Malus apple trees is their scarcity in Israel today. It is argued that these apple trees will not grow in that hot and dry climate, whereas the apricot and quince will.

There are a few answers to this. First, apple trees as we know them were seriously cultivated by the Romans in Israel. They believed that there were health benefits in the apple, and so they cultivated them when they conquered Israel. This establishes the fact that Malus apple trees can grow in the climate of Israel.

Also, within the last twenty years, Malus apple trees have again been cultivated in Israel. They flourish in the north near Lebanon. But small, sweet apple trees are also now growing in Gaza, an arid area, but close to the Mediterranean Sea. Through proper cultivation and grafting knowledge, these apple trees can be made to grow throughout most of Israel.

In exploring the ability of Biblical Israel to grow apple trees, it is interesting to note that the apple trees in Gaza are located on the thirtieth parallel. This thirtieth parallel, followed straight west from Gaza leads right to the city of Bethtappuah, the "House of Apples," about halfway between Gaza and the Dead Sea. There, sitting on the parallel where apples have been proven to grow, is one of the cities named for its abundance of apples, lending great credibility to the view that the Biblical apple was the Malus apple.

On the whole, there is no reason to doubt that the Biblical apple tree was a Malus apple tree. The Bible references don't refute this position and, in fact, strengthen it. Also scientific evidence has not refuted it, but instead strengthened it, as in the cultivation of apple trees on the thirtieth parallel. Even the modern Hebrew name for the Malus apple is the same as that for the Biblical apple, tappuah, and the Arabic name is very similar, taphuah. Again, it seems that the translators got it right when they translated this word as "apple."

Another point not often considered is the relatively small amount of water required by fruit trees. Fruit trees flourish on the tops of hills where the ground is not saturated with water. They don't do as well in valleys where the water is more abundant. It is nut trees that flourish in the valleys. They like what arborists call "wet feet," having their trunks rooted in well-watered ground.

This at first seems a little odd, that juicy fruit needs less water to grow than dry nuts. But it is the case. While it must be admitted that fruit trees cannot live without any water, they do need good drainage. For this reason, they prefer drier sites even than nut trees.

One more point to consider is the change in Israel's climate. It has not always been as arid as it is now. When the Israelites were first promised the land, it was described as *"a land flowing with milk and honey." (Exodus 3:8)*

When the spies entered the land in roughly the 1500s B.C., they found it just as had been promised. *"And they came unto the brook of Eshcol, and cut down from thence a branch with one cluster of grapes, and they bare it between two upon a staff; and they brought of the pomegranates, and of the figs." (Numbers 13:23)* It took two of these spies to carry one cluster of grapes. Either the grapes were very large, or the cluster had a lot of grapes on it. Either way, it was a land that could produce fruit in great abundance. The spies echoed the Lord's promise, saying, *"surely it floweth with milk and honey; and this is the fruit of it." (Numbers 13:27)*

Even in Solomon's time, around 1000 B.C., the climate was conducive to fruit-bearing trees. Solomon was known for his cultivation of gardens and trees. He said, *"I made me gardens and orchards, and I planted trees in them of all kind of fruits." (Ecclesiastes 2:5)* Solomon was no novice, but an expert in botany. It was he who mentioned the apple tree several times in the Song of Solomon. He knew what he was talking about, and no doubt this was one of the trees that he cultivated in this lush become discontented and had asked for a king in order to be like all the other nations about them. In answer to their complaint, God had given them Saul as the first king of Israel, who proved to be a very wicked king. This is a sobering lesson for us. We must be careful what we ask for, because God just might give it to us, even if it is not good for us.

As Saul's reign progressed, he began to manifest his rebellious and disobedient heart. He grew impatient to go to battle against the Philistines and refused to wait for Samuel the priest to offer the sacrifice beforehand. He instead offered it himself, defying God's established order. He later disobeyed God's command to destroy the Amalekites completely.

For this, God rejected him as king and had Samuel anoint David as the next king. However the transfer was not immediate, and Saul fell deeper into sin. Knowing David was set to inherit the throne, Saul tried to kill him and very intensely sought his life, so that one of his own sons would be his successor.

The description of this tree as a **_green_** bay tree gives further insight into the life of Saul. This word indicates that the tree is young. But as is so often the case, though it is young and strong right now, it has failed to prepare for the day when it will be old and feeble.

Young trees have soft wood that is very flexible, but counter-intuitively, this makes them very strong. They can withstand high winds and much bending and twisting without breaking. Thus like young people, young trees are stronger than older ones. But when trees get old, though their wood becomes harder, they actually become weaker. They can no longer bend under stress, and so they break easily. And so, like elderly people, older trees are hard and "set in their ways," but they are also fragile and prone to breaking.

Thus this young tree, heedless of these coming days, spends all its energy upon his present growth and power. For the word "green" reveals not only its youth, but also something of its growth. Very green trees are trees that grow fast. They especially grow fast when they are native, as the word "bay" means, growing in their own native soil. This fast growth and greenness reveals that this tree spends all its energy on growth and leaves, and not on defense. Like this tree, Saul was out to grow his own power, and not to seek the good of Israel.

But if a tree grows fast, it also dies fast. It's sap is mostly water, and it lacks substance. Therefore it can very quickly wither away. We see an example of this in grass during the spring. Grass grows very fast at this time of the year. But 75 percent of its blade is mere water. Only 25 percent is the actual substance that makes up grass. For this reason, we can cut grass on one day, but by the next, we can hardly see the cut blades lying around. They have shriveled up and withered almost into nothing.

In the same way, when the wicked grows fast and spreads himself like a green bay tree while he is young, he lacks substance. He will die and wither away quickly, and will be unable to be found. He will become hard and brittle, and will therefore break and be destroyed. There will be no trace of him left. David said, *"Yet he passed away, and, lo, he was not: yea, I sought him, but he could not be found." (Psalm 37:36)* Such is the sudden and swift destruction of the fast-growing, powerful wicked man who seeks his own power and authority.

In order that we might avoid this destruction, let us learn the lesson that David learned and has taught to us through the bay tree. *"Wait on the LORD, and keep his way, and he shall exalt thee to inherit the land: when the wicked are cut off, thou shalt see it." (Psalm 37:34)* reddish-brown pitch, mostly red, which, especially when it is dry, looks like blood. This blood-colored pitch would have been both within and without the ark, picturing the need for Christ's blood to cover us both within and without. Its sweet fragrance also would have pictured the truth that Christ's sacrifice was a sweet savor to God.

Indeed, the root of camphire, *kaphar*, is the same word translated "atonement" in the Bible. With a slightly different spelling in English, but the same word in Hebrew, it is the word *kippur* in Yom Kippur, the Day of Atonement. While the atonement of bulls and goats was only temporary, the atonement of Christ is permanent and is our sins' only true covering. While the red camphire may give us temporary beauty and temporary perfume, it is only the blood of Jesus Christ that can permanently remove the stain of our sins and give us His Own true beauty and sweet fragrance in their place.

9
The Cedar Tree

The cedar tree is used abundantly in the Bible, often as a symbol. The Bible repeatedly uses earthly symbols to teach heavenly truths. For example, the imperfect sacrifices of bulls and goats in the Old Testament are a type of Christ's perfect sacrifice in the New Testament. Though animal sacrifice cannot impart real redemption, it does serve as a picture of redemption. Another example is the parable of the sower. The four kinds of soil picture the four different ways that people receive truth. The list could go on an on with the parable of the lost sheep and the numerous other parables and symbols in the Bible.

The three books of the Bible that use symbols the most are Ezekiel, Daniel, and John. And of these, the one with the highest concentration of plant or agricultural symbols is Ezekiel. Indeed most of what we know about plants in the Bible comes from this middle section of the Bible. The first few books of the Old Testament and the entire New Testament are both pretty quiet about plants. This leaves most Biblical references to plants in the middle part of the Bible.

Of these plants, one of the most prominent and oft-mentioned is the cedar. It is used for many things in the Bible, and we have already come across it several times. In this chapter, we will try to hit the high points and catalog its main characteristics and uses, as well as bring out and clarify some of the references to it that may be obscure or confusing.

The Hebrew name for the cedar is *erez* (אֶרֶז), the Greek name is *kidros* (κιδρος), and the Latin name is *Cedrus*. As we can see, there are similarities between all of these names.

Of course, the first thing that comes to mind when we think of the cedar is the cedar of Lebanon. We will also later learn from Ezekiel 31, which we will study in chapter 26 of this book, that the cedar is a stately tree and has long branches. Song of Solomon speaks of the cedar tree as a fragrant tree. This is because there is a lot of tannin in cedar wood. I Kings and II Chronicles record that cedar wood was used in construction of the temple and palace, and therefore it would have to be durable wood.

The placement of the cedar wood in the temple and palace gives us further understanding of its nature. It was never used for structural floor and ceiling joists, but was used mainly in the walls and also to cover the structural boards of the floor and ceiling, which were made of fir.

This is because cedar wood is very strong vertically, but it is flexible horizontally. This can be simply illustrated by a drinking straw. When we push on a straw from the top, it remains firm and is difficult to bend. But when we push on the same straw from the side, it bends easily. For this reason, cedar worked well as a wall stud. It stood strong against the downward pressure of the ceiling. But it did not work as well as a floor or ceiling joist. The pressure that would have come on its side in this position would have caused it to sag. Only horizontally strong wood, like fir, can be used for this function. Still cedar wood was needed to cover not only the walls, but also the structural floor and ceiling beams because of its ability to resist rot and decay.

This would make a perfect combination that would make the temple durable and long lasting. The fir joists would give stability for the priests to walk and offer sacrifices, and the cedar covering and cedar walls would provide protection from decay and insect infestation. All of this would also be overlaid with gold, giving further protection, but also beauty and glory to this house of God's presence.

There have been many other uses for the cedar throughout history. It has been used for carving and for building ship masts. Zerubbabel, like Solomon in his first temple, also used cedar for building. Cedar sawdust has even been found as packing in the coffins of Pharaohs of Egypt.

The cedar grows best at high altitudes, around 3,500 feet or even higher. This helps us to understand better the image in Ezekiel 17 of the highest branch of the high cedar, set upon a high and eminent mountain. *"Thus saith the Lord GOD; I will also take of the highest branch of the high cedar, and will set it; I will crop off from the top of his young twigs a tender one, and will plant it upon an high mountain and eminent." (Ezekiel 17:22)*

The cedar also has a very wide branch spread. A 90-foot tall cedar tree can have 50-foot long branches. With branches extending 50 feet in all directions, this tree would have a total diameter of 100 feet, making it wider than it is tall. Because there is nothing going from the tree to the ground for a 100-foot swath, it looks like the branches are floating in midair. This is an impressive sight and demonstrates the great strength of the cedar, for not only are these branches long, but they are also very dense. Only very strong wood could hold such weight out from the trunk at such a distance. Its horizontal flexibility contributes to this ability, making it capable of bending under the weight without breaking rigidly.

This wide spread of strong and dense branches provides a large area of shade and protection under which many people can sit. It also provides abundant lodging for birds in it branches. For these reasons, the Bible likens the great kings Pharaoh and Nebuchadnezzar to cedar trees, because many nations and people could sit under their shadow and rest in their branches.

For those of us who are familiar with the redcedar of the United States, we must clarify that we are not talking about this tree. The redcedar is not large and spreading at all, but instead is very small. This is because it is not a true cedar, but really a juniper. Its Latin name is *Juniperus*, whereas the Latin name of the Biblical cedar is *Cedrus*. It is this Biblical cedar that has the great spread of branches.

What kind of cedar, then, is the Biblical cedar? Most believe and I am convinced it is the kind of cedar known today as the "cedar of Lebanon." This is interesting because for all of the times this cedar is mentioned in the Bible, the phrase "of Lebanon" is used only five times, once in Judges, twice in Psalms, and twice in Isaiah. We quote here the one in Psalm 29, *"The voice of the LORD breaketh the cedars; yea, the LORD breaketh the cedars of Lebanon." (Psalm 29:5)* Aside from these five places, all others refer to the cedar as in, out of, or from Lebanon. Yet the name "cedar of Lebanon" has stuck, and is today the name of what is probably the Biblical cedar.

Some commentaries still say that the Biblical cedar is really a pine or a fir. But though these are all evergreens and may look very similar, these terms are very specific and cannot be used interchangeably. True cedar cones and cedar wood are distinct and different from the cones and wood of pine and fir trees.

We must realize that the words of the Bible are specific because the Lord, their author, is specific. He gave specific instructions for carrying the ark of the covenant, that it was to be borne on the shoulders of the priests. When it was instead transported on an ox cart and touched by Uzza who sought to steady it, even this sincere man was struck dead for violating God's specific instructions. God's words are specific, and when He says cedar, He means cedar.

The cedar tree was highly prized also in Solomon's day, and was a status symbol. It was extremely rare in Israel itself, which is why it had to be imported from Lebanon. Yet the Bible says they imported *"cedar trees in abundance: for the Zidonians and they of Tyre brought much cedar wood to David." (I Chronicles 22:4)*

This highly regarded and highly prized cedar tree is a picture of Jesus Christ in many ways. It is impervious to insects and disease, just as Christ is impervious to sin and to all that would mar or harm Him.

The large, spreading branches of the cedar illustrate Christ's care for all of His Own. Just as the cedar branch is always there for the many birds and people who take refuge in and under it, Christ is always with us. Though we may not always feel prosperous, we can rest in the fact that He is with us, sheltering and sustaining us.

Also the cedar's fragrant smell points to Christ, whose sweet savor on our behalf is spoken of in Ephesians 5. *"Christ also hath loved us, and hath given himself for us an offering and a sacrifice to God for a sweetsmelling savour." (Ephesians 5:2)*

The cedar is also an evergreen. Evergreens are the oldest living trees on earth. As trees continue to be discovered, the one designated the oldest tree is constantly changing. It used to be a husa tree, but now another evergreen has been discovered that is even older. Still the oldest tree is always an evergreen. This reminds us of Christ Who is everlasting and the *"Ancient of Days" (Daniel 7:9)* and both *"Alpha and Omega, the beginning and the end." (Revelation 22:13)* But also like Christ, the evergreen, even with all of its age, always remains green and fresh. In the same way, Christ has existed from eternity past, but He is *"the same yesterday, and to day, and for ever." (Hebrews 13:8)* He is just as alive, vibrant, and powerful as He has always been.

Of course, the cedar in the temple overlaid with gold is also a picture of Jesus Christ. He is strong and prevents decay, and also He shines with the glory of pure gold.

An interesting and tragic comment is made concerning Shallum, the son of Josiah, in Jeremiah 22. Apparently Shallum had wanted his house to be better than the temple. He received this pronouncement of judgment. *"Woe unto him that buildeth his house by unrighteousness, and his chambers by wrong; that useth his neighbour's service without wages, and giveth him not for his work; That saith, I will build me a wide house and large chambers, and cutteth him out windows; and it is cieled with cedar, and painted with vermilion." (Jeremiah 22:13-14)*

Shallum was not content with cedar alone. He went beyond that and used **_painted_** cedar, vermilion being a reddish dye. He tried to give himself greater glory than God's cedar temple and than God Himself. And he did this all at the expense of the people he ruled. Shallum sought his own glory to the exclusion and even the destruction of all others. For this pride and selfishness, God called his a house built *"by unrighteousness." (Jeremiah 22:13)*

From this sobering account of cedar, we should be warned not to seek to usurp the glory of Christ. He is the highly prized cedar, and He alone. Let us glorify Him as such and not seek to lift up ourselves.

The cedar tree also appears in Ezekiel 27 as being used to make masts for ships. This chapter beings with a lamentation over Tyre, or Tyrus. *"The word of the LORD came again unto me, saying, Now, thou son of man, take up a lamentation for Tyrus; And say unto Tyrus, O thou that art situate at the entry of the sea, which art a merchant of the people for many isles, Thus saith the Lord GOD; O Tyrus, thou hast said, I am of perfect beauty." (Ezekiel 27:1-3)*

Tyre was one of the chief cities of the kingdom of Lebanon, and so the kingdom was known as the kingdom both of Tyre and of Lebanon. This kingdom abounded with cedars. It was from here that Hiram, king of Tyre, exported the great abundance of cedars and sent them to Israel for use in the temple. I Kings says that, after this exportation, cedars were so abundant in Israel that they were *"as the sycomore trees that are in the vale, for abundance." (I Kings 10:27)* This is quite a statement, for sycamore trees were very common and abundant in Israel.

This means that in Lebanon, the cedar trees would have had to have been even more abundant than this. The great number of cedars in Israel was only what Lebanon exported, not its total number of cedars. Still today, Lebanon remains famous for cedars. It even has a cedar grove that is called the "Cedars of the Lord God."

The Tyrians, the people of Tyre, were also famous for something else. The next verse of Ezekiel 27 says, *"Thy borders are in the midst of the seas, thy builders have perfected thy beauty." (Ezekiel 27:4)* Tyre's borders were *"in the midst of the seas."* Tyre was a coastal city, bordering right upon the sea. For this reason, the Tyrians, called also Phoenicians, are known in secular history as the first people to build ships. This has recently been questioned, but even if they were not the very first, we do know they at least improved on ship building, and they were well known for this industry.

The next two verses go into detail describing the different woods used for the different parts of the ship. *"They have made all thy ship boards of fir trees of Senir: they have taken cedars from Lebanon to make masts for thee. Of the oaks of Bashan have they made thine oars; the company of the Ashurites have made thy benches of ivory, brought out of the isles of Chittim." (Ezekiel 27:5-6)*

Why were each of these specific kinds of wood used for these different parts of the ship? Why was cedar chosen for the masts over fir and oak? We will fully cover the fir and the oak later when we get to their own chapters, but for now we look at them only as they compare to the cedar for use in shipbuilding.

One reason the cedar was chosen is its high strength to weight ratio. When building a ship, weight, not size, is the important factor. The ship must be light in order to float high in the water. Shipbuilders always strive for the greatest strength with the least weight.

With this in mind, wood is a good substance for shipbuilding because it has a greater strength to weight ratio than that of such substances as steel or plastic. The reason for this is that for every ounce of weight, you can have more wood than you can these other substances. For example, a thin plate of sheet metal weighs the same as a thick board of wood, but the thick board is stronger.

The same principle applies within different kinds of wood itself. Soft wood has a greater strength to weight ratio than hard wood. One pound of soft pine, for example, contains more woody substance than one pound of hard oak. Because it has more substance, the pound of pine is stronger than the pound of oak.

For this reason, cedar, a soft wood, is a good choice for building the masts here in Ezekiel. Shipbuilders want masts, like the ship itself, to be strong, but not heavy. Heavy masts make the ship sit low in the water, and they are also difficult to raise. The cedar strikes a balance. It is strong, but it is not a heavy wood like oak.

Were the ship made all of oak, it would sit very low in the water due to its weight. Sometimes this is desirable for the sake of strength and defense. For example, the English navy made many of their ships of oak. This makes for a very strong ship, and if it is white oak, it is also impervious to water. The only problem is its great weight.

To avoid this problem, the fir was used to build the main body of the ship because it, like cedar, is also a medium soft wood. But fir has other problems that make it less desirable for use as a mast. One is that it has many knots. Wherever there is a knot, the wood is weak at that point. A mast, standing straight up in the air and supporting the weight of a large sail, needs to be very strong. So fir wood, full of knots, would not serve the purpose well. Cedar wood, on the other hand, has less knots than fir, and so would make for a strong mast, stronger than either fir or oak.

The cedar is also resistant to rot. All this combined makes it the perfect solution for use as a mast. It is resistant to rot, relatively free of knots, and very strong for its light weight. It is also a very straight and tall wood, making it ideal to function as a mast. Still today the best wood for making a mast is Port-Orford-Cedar, though it is becoming hard to obtain.

We see again God's perfection and detail in all that He says and does. The different uses of the oak, fir, and cedar in the construction of these ships were not chosen arbitrarily. They were chosen with a perfect understanding of the nature of each wood and a perfect matching of it to the task it was to perform.

In the same way, God makes every Christian different in order to fit into His overall purpose. Each of us has our own strengths and weaknesses. But these all work together to complement one another and to make each of us perfectly suited for our unique task within the body of Christ. All the while, God sees the entire body and sees how the labors of individual Christians combine to accomplish His overall purpose.

Another interesting and somewhat confusing reference to the cedar is in Numbers 24. Here Balaam had been trying to curse Israel as they traveled on their journey to the promised land, but he was forbidden by the Lord and could only bless them in Numbers 24:1-2.

Once more Balaam would speak. Again, against his will, it would be a blessing, put in his mouth by the Spirit of God. *"And he took up his parable, and said, Balaam the son of Beor hath said, and the man whose eyes are open hath said: He hath said, which heard the words of God, which saw the vision of the Almighty, falling into a trance, but having his eyes open: How goodly are thy tents, O Jacob, and thy tabernacles, O Israel! As the valleys are they spread forth, as gardens by the river's side, as the trees of lign aloes which the LORD hath planted, and as cedar trees beside the waters." (Numbers 24:3-6)*

This verse has often been misunderstood and questioned because it describes cedar trees as *"beside the waters."* Many commentaries say that this cannot be the true Biblical cedar, the cedar of Lebanon, because it doesn't grow beside water. This is true, but to conclude that the verse is talking about a different kind of cedar is to miss the point of the verse entirely.

The context of the verse must be kept in mind. Balaam saw Israel while they were still traveling, wandering in the wilderness. He saw them outside of the promised land, not in a permanent home, but in tents, in a place where they didn't belong. For this reason, he compared them to trees that are where they don't belong, *"cedar trees beside the waters."* The key here is the phrase *"which the LORD hath planted."* God can plant trees wherever He likes and can sustain them and make them grow miraculously where they won't naturally. In the same way, He can sustain His people even outside of the land in which they belong.

In only about two short years, this sustenance would come to fruition. God would at last bring His people into the promised land and plant them in the place where they belonged.

One final use of the cedar that we will look at is in Leviticus 14. Here we find the cedar used twice in the cleansing of leprosy. In the early part of the chapter, it is used to cleanse a person of the disease. *"Then shall the priest command to take for him that is to be cleansed two birds alive and clean, and cedar wood, and scarlet, and hyssop: And the priest shall command that one of the birds be killed in an earthen vessel over running water: As for the living bird, he shall take it, and the cedar wood, and the scarlet, and the hyssop, and shall dip them and the living bird in the blood of the bird that was killed over the running water."* (Leviticus 14:4-6)

In the latter part of the chapter, the cedar is again used, this time to cleanse a house of the disease of leprosy. *"And he shall take to cleanse the house two birds, and cedar wood, and scarlet, and hyssop: And he shall kill the one of the birds in an earthen vessel over running water: And he shall take the cedar wood, and the hyssop, and the scarlet, and the living bird, and dip them in the blood of the slain bird, and in the running water, and sprinkle the house seven times."* (Leviticus 14:49-51)

Why is the cedar used for this? Why not use the wood from any old tree, an oak, a fir, or a fig? It is for the sake of

Matthew 13 records a reference to Christ as the carpenter's son. In this passage, Christ has just returned to Nazareth, the village in which He was raised. There He begins to teach the Nazarenes in the synagogue. But they ask, *"Is not this the carpenter's son? is not his mother called Mary? and his brethren, James, and Joses, and Simon, and Judas?"* (Matthew 13:55) They are trying to demean Christ, saying that He is "just a carpenter's son," and questioning, "Why is He teaching us?"

This same account is recorded in Mark, where the Nazarenes say, *"Is not this the carpenter, the son of Mary, the brother of James, and Joses, and of Juda, and Simon? and are not his sisters here with us? And they were offended at him."* (Mark 6:3) These men are offended because they view Christ as "just a carpenter." They are too proud to be instructed by a man that they view as a lowly carpenter.

But in reality, though they do not know it, Jesus Christ is a far better carpenter than any other. He makes things that will not burn, but that are permanent. He has the power even to make these men themselves into new creatures, having everlasting life.

Christ learned His earthly carpentry skills from His earthly father Joseph, just as these men supposed. But He learned His heavenly carpentry skills from His Father in Heaven. With these skills, He is the true "Carpenter's Son." He has the power to make each of us into a piece of workmanship that will never burn and never perish.

This question *"Is not this the carpenter's son?"* is far more than a simple degrading of Christ. It is a fundamental questioning of His deity. Years later when Christ stood before the Sanhedrin men were still questioning His deity. *"Then said they all, Art thou then the Son of God? And he said unto them, Ye say that I am."* (Luke 22:70) They were still thinking of Him only as the son of Joseph.

Satan had also questioned Christ's deity when he tempted Him at the opening of His public ministry. *"And when the tempter came to him, he said, If thou be the Son of God, command that these stones be made bread."* (Matthew 4:3)

Even today men continue to question the deity of Jesus Christ. A famous politician was recently asked, if Jesus Christ were here today, what one question would he like to ask Him? The politician said his question would be, "Are you really the Son of God?" Today, two thousand years after Christ walked on earth, men are still asking this same question. They are questioning His deity simply because they refuse to believe the Bible, *"the record that God gave of his Son."* (I John 5:10)

Jesus Christ in a very real sense is both the son of Joseph the carpenter and the Son of God the Master Craftsman. Even today He is still a carpenter even in the physical sense. Hebrews speaks of Abraham looking *"for a city which hath foundations, whose builder and maker is God."* (Hebrews 11:10) Jesus Christ is building a city in heaven even now. He is making mansions for us.

The mansions that He builds are unique from the buildings here on earth. Here our earthly buildings eventually rot and decay. But Jesus Christ is building a heavenly and eternal city. He said to His disciples, *"Let not your heart be troubled: ye believe in God, believe also in me. In my Father's house are many mansions: if it were not so, I would have told you. I go to prepare a place for you."* (John 14:1-2)

Though Jesus learned earthly carpentry skills from Joseph, He learned these heavenly carpentry skills from His heavenly Father, and He uses His skills to build us an eternal mansion. When Jesus left the earth, He did not leave behind His occupation. He was a carpenter then, and He is still a carpenter today. As such, He also fulfills the carpenter's role as an engraver. As He fashions the heavenly city, He carefully engraves it so that it is not only functional, but also beautiful.

As the Master Carpenter, Jesus Christ continues His carpentry work today, in our physical lives, in our spiritual lives, and in our eternal home in heaven.

12
Cinnamon

Cinnamon is mentioned four times in the Bible. Though it is small, we examine it here because it is indeed a woody plant. Some may argue that it is really a shrub and not a true tree, but most of the time the only difference between a shrub and a tree is how it is trimmed. So since the cinnamon is woody, we include it in our study.

There is no controversy over what the cinnamon plant is, like there is for so many of the Biblical trees. The Hebrew name *qinnamown* (קִנָּמוֹן), the Greek *kinamomon* (κιννάμωμον), the Latin *Cinnamomum*, and the English cinnamon are so clear and so similar that none can deny that the Biblical cinnamon is the cinnamon we know today.

We will notice as we study the cinnamon that its wood is the important part of this plant, as opposed to its fruit. The fruit of the cinnamon is never mentioned, but its wood and the byproducts that come from its wood are the important and usable parts of this plant.

Cinnamon first appears in Exodus 30 as a spice used in the holy anointing oil of the tabernacle. *"Take thou also unto thee principal spices, of pure myrrh five hundred shekels, and of sweet cinnamon half so much, even two hundred and fifty shekels, and of sweet calamus two hundred and fifty shekels." (Exodus 30:23)*

Notice here that this cinnamon is called *"sweet cinnamon."* There are two different kinds of cinnamon, white cinnamon and dark cinnamon. It is the white cinnamon, also called Ceylon cinnamon, that can be prepared into a sweet spice, whereas the dark cinnamon can be prepared in the same way, but will remain bitter.

This preparation process for making a white cinnamon tree into sweet cinnamon spice begins with peeling the bark from the branches. The cinnamon shrub is related both to the sassafras and to the laurel. Like these, its entire plant is aromatic, but it is the bark that contains the concentrated substance that can be turned into the cinnamon spice. Once this bark is peeled off, it is then allowed to dry. As the moisture leaves, the bark curls up, and only then does it become a sweet spice.

This spice is used in two very different and contrasting ways in the Bible. First in Proverbs, cinnamon is used by the harlot to attract the young man void of understanding. She says to him, *"I have perfumed my bed with myrrh, aloes, and cinnamon."* *(Proverbs 7:17)* Here cinnamon pictures the false attraction of the adulterous woman and an appeal to sensual lust. It demonstrates a very vile use of the cinnamon.

However in Song of Solomon, this same plant is used by the very opposite of the harlot, the beloved one. Found in the garden of the beloved are *"spikenard and saffron; calamus and cinnamon, with all trees of frankincense; myrrh and aloes, with all the chief spices." (Song of Solomon 4:14)* Here the cinnamon pictures the sweetness of true love between a man and his wife. It demonstrates a very lovely and sweet use of the cinnamon, in contrast to the vile use in Proverbs.

Four important contrasts can be drawn between these two passages. First, there is the contrast we have noted between the strange woman and the truly beloved. Second, there is the contrast between the ordinary spices of Proverbs and the addition of *"chief spices"* in Song of Solomon. Third, in Proverbs a mere woman is speaking, not even a lawful bride, but in Song of Solomon the legitimate bridegroom is speaking. And fourth, the strange woman leads to the *"chambers of death" (Proverbs 7:27)*, whereas the beloved leads to *"living waters." (Song of Solomon 4:15)*

The last occurrence of the cinnamon spice in the Bible is in Revelation 18. Here Babylon has just fallen, and the merchants of the last days are weeping over the loss of their valuable trade in which they have trafficked. They lament over the loss of *"cinnamon, and odours, and ointments, and frankincense, and wine, and oil, and fine flour, and wheat, and beasts, and sheep, and horses, and chariots, and slaves, and souls of men." (Revelation 18:13)* Here we find cinnamon ranked with ointments, frankincense, and wine. It is a very valuable and lucrative spice whose loss they mourn.

It is interesting to note that in this chapter of Revelation, wood is always spoken of as a rarity. Not only is the rare cinnamon named in verse 13, but also in verse 12 *"thyine wood,"* another rare wood, is named, along with the general term *"most precious wood."*

I believe this scarcity of wood is due to the devastating effect that the vial judgments of chapter 16 have upon the trees of the earth. The second vial is poured on the sea and turns it into blood, the third vial is poured on the rivers and turns them into blood, and the fourth vial is poured on the sun and scorches the earth with fire. All of this has a disastrous effect upon the trees, depriving them of life-giving water, and burning them up with scorching heat. So by chapter 18, wood is a very rare commodity, in which cinnamon is included.

But exceeding the cinnamon as a spice with its lucrative enterprise and its picture of the beloved, is the cinnamon as an oil. And it is with this that we conclude our look at the cinnamon, by returning to its first and greatest Biblical use, its only Biblical use as an oil, its use in the holy oil of the tabernacle.

One cinnamon plant can yield up to five or six quarts of cinnamon oil, thus furnishing an abundant supply for this its supreme use. As God commanded, this oil was used to anoint only the most holy objects, the tabernacle, the items within it, and the priests, thus marking these as holy and set apart unto the Lord.

13
The Cypress Tree

In other translations of the Bible, the word "cypress" may appear frequently, depending on the opinions of the translators. But in the King James Version, it appears only once. This verse says, *"He heweth him down cedars, and taketh the cypress and the oak, which he strengtheneth for himself among the trees of the forest: he planteth an ash, and the rain doth nourish it." (Isaiah 44:14)* Here "cypress" translates the Hebrew word *tirzah* (תרזה), which, like the English word, is found only here.

Many translations say that this tree is really a fir, but the fir is in the *Pinaceae* family instead of the *Cupressaceae* family of the cypress. Also the Hebrew word translated fir is *barowsh* (ברוש). If these were the same tree, why would they be called by two different names, both in Hebrew and in English? Also cypress wood can be hard and, as such, fits the description of its use in Isaiah 44, but the fir is not considered hard and so wouldn't work well for use in making an idol.

Some translations also say the cypress is a kind of oak. But if this were the case, the verse would read, "He . . . taketh the oak and the oak." This would be redundant and absurd.

Other translations propose that cypress wood is the wood referred to in Genesis as gopher wood, of which the ark was made. The cypress seems ideal for the ark because it is very resistant to rot and can last a long time. But we must remember that the ark needed to float for only about one year. Most any wood will last that long. You can even buy a two-by-four, throw it in your backyard, and it will last at least one year.

I personally don't think God intended the ark to last a long time. For one thing, it didn't need to last long in order to accomplish its task. And for another, if it had remained intact long after its task was over, it may have become a temptation to worship it, rather than God Himself, just as the brazen serpent became. This serpent, originally a good object for the healing of the people, eventually had to be destroyed when the people worshipped it, rather than God. The same is true of the wood of which the cross was made. While the wood served its purpose for its time, if we still had it, we would be in danger of worshipping it, rather than the Christ Who died on the cross.

I don't believe the cypress to be the gopher wood used in the ark. My opinion of gopher wood will be covered in chapter 21, that it is not a kind of tree, but rather a process. The Biblical cypress I take to be just that, a true cypress. This tree grows abundantly in the mountains of Taurus and Hermon.

We now look at how the cypress is used in the Bible. The reference in Isaiah is referring to the making of idols. *"He heweth him down cedars, and taketh the cypress and the oak, which he strengtheneth for himself among the trees of the forest: he planteth an ash, and the rain doth nourish it. Then shall it be for a man to burn: for he will take thereof, and warm himself; yea, he kindleth it, and baketh bread; yea, he maketh a god, and worshippeth it; he maketh it a graven image, and falleth down thereto."* *(Isaiah 44:14-15)* All the trees named here were used both for firewood and for making idols. They would be carved into figures, such as bulls or goats, and then worshipped.

The first three trees named, the cedar, the cypress, and the oak, are ideally suited for such use. They can be carved out, and their wood is not known to rot. This resistance to rot was desirable in idols as Isaiah 40 reveals, *"The workman melteth a graven image, and the goldsmith spreadeth it over with gold, and casteth silver chains. He that is so impoverished that he hath no oblation chooseth a tree that will not rot; he seeketh unto him a cunning workman to prepare a graven image, that shall not be moved."* *(Isaiah 40:19-20)* If the poor man could not afford gold and silver, he would choose the next best thing, a durable wood. The images made from these rot-resistant trees can be kept around a long time, as opposed to the softer trees such as poplar or fir. That is why the cypress was included in this list.

The cypress, mentioned in the Bible only as being used for idol worship, teaches a tragic lesson. Its hardness and resistance to rot would have made it useful for many things. Yet its potential was squandered, and its good characteristics were wasted on idolatry. Let us learn from this tree that all good things that we have are given to us from God. They are not ours to use as we will according to our own pleasure. They must be used only for God's glory, or else they are wasted, and rather than accomplishing the good that God intended, they turn to our harm and detriment. Let this not be said of the good things God has given us. Rather, let us use all for His glory.

14
The Ebony

The ebony is another tree that is mentioned only once in the Bible. It is found in Ezekiel 27 in a whole list of merchandise for which the Tyrians traded. Ebony is listed as one of the substances they obtained from Dedan. *"The men of Dedan were thy merchants; many isles were the merchandise of thine hand: they brought thee for a present horns of ivory and ebony." (Ezekiel 27:15)*

Nowhere in the Bible are so many nations named together as are named here in Ezekiel 27. We must remember that this passage was addressed to the Tyrians, or Phoenicians. They were the leaders in ship building and sea trade. They had contact and merchant trade with many nations.

Ebony is a very dark wood and a very hard wood. It seems that it came from what is today India. Coming from this foreign land, it is exotic, rare, valuable, and expensive. This shows what a huge and powerful navy the Tyrians had, great enough to travel this long distance for such an expensive item.

The Hebrew word for ebony is *hoben* (הבן). The Latin term is *Diospyros ebenum*. It means black. Ebony wood has jet black heartwood and can have yellowish brown streaks running through it. The unique black color of this wood adds to its value. In fact, ebony is so black that it was at one time used to make black piano keys.

What does the ebony teach us? It teaches us the value of other brothers and sisters in Christ. The nation of Tyre could not sustain itself on its own. It needed trade with other countries. In fact, one of its greatest treasures was found in one of the most distant countries. Like Tyre, no nation is self-sufficient. None has all available resources within its own borders. Some are good for mining, others for agriculture. In order to live a full and healthy life, nations must trade with one another.

In the same way, no person is self-sufficient. God has made it so that we all need one another. The weaknesses of one are balanced by the strengths of another. Let us learn to befriend and invest in even the most distant and removed, for they may be the very ones who in turn contribute the most valuable treasures to our lives.

15
The Elm Tree

The only place the elm is mentioned in the Bible is Hosea 4:13. Here God is describing the idolatry of His people. He says of them, *"They sacrifice upon the tops of the mountains, and burn incense upon the hills, under oaks and poplars and elms, because the shadow thereof is good: therefore your daughters shall commit whoredom, and your spouses shall commit adultery." (Hosea 4:13)*

The Hebrew word for elm here is *elah* (אלה). There are several suggestions for how this word should be translated. The word for oak in the same verse is very similar, *allon* (אלון), and some say that both refer to an oak tree. In my opinion, because they are used in the same verse and the words, though similar, are not identical, they are two different trees. Otherwise the verse would read "under oaks and poplars and oaks," with unnecessary redundancy.

It is true that elsewhere in the Bible the word *elah* (אלה) is translated "oak." But the word *elah* simply means "strong." Two different trees can both be strong, but be different trees. I believe this to be the case here. Both elms are oaks are strong trees, and so I believe they are both accurate translations of the Hebrew word *elah* in their own verses.

Another tree that has been suggested and is even used in many Bible translations to translate *elah* or a related word, *elim*, is the terebinth tree. But the terebinth is never named in the King James Version of the Bible and has properties unlike the elm.

One other tree that must be included in this discussion is the teil tree. It is named in Isaiah 6:13, *"as a teil tree." (Isaiah 6:13)* But the Hebrew word it translates is the very same word translated elm in Hosea, *elah*, meaning "strong tree." For this reason, many say that the Biblical elm is the same as the terebinth, which is the same as the teil, that all three are the same tree. However, I believe they are all translated correctly, and that reasons.

61

The terebinth tree is what we classify under the Latin term *Pistacia terebinthus*. It is a pistachio tree. On the other hand, the elm is not in the genus *Pistacia*, but in the genus *Ulmus*. This is important because of the difference in size between these two. The terebinth and all pistachio trees are not as tall as elm trees. Also, the terebinth tree has pinnate leaves and red berries, both of which are not features of the elm tree.

In the Biblical reference, we find that the children of Israel are burning incense under elms, poplars, and oaks *"because the shadow thereof is good." (Hosea 4:13)* These cannot be small trees, but rather have to be pretty tall. They have a good shadow and are large enough for incense to be burned underneath them. We know that the oak and the poplar are both tall trees, and it would seem natural that the other tree used for the same purpose and, like them, described as having a good shadow and incense burned underneath it, would also be tall.

While some pistachio trees can be tall when they are in the right soil with the right temperature and the right climate, they are not tall as a general rule. But elm trees are consistently tall. They better fit the description of Hosea.

Some people argue that the elm cannot grow in Israel. But this is not entirely true. While common North American elms don't grow there, some elms do, such as the white elm, also called the Russian elm, and the rock elm. These are all in the *Ulmus* family. They have leaves and a general appearance that are similar to all elms.

We also don't know exactly what the climate was like in Israel during the time of Hosea. It was certainly better than it is now. Solomon's gardens flourished only a century or two earlier, and it wasn't until after Hosea, during the days of Haggai, that God cursed the ground with dryness and barrenness as a result of Israel's sin. In that day, most any tree would have thrived in Israel.

The elm tree has proven useful throughout history. One reason is its great ability to resist rot. Even when its wood is put into water, it has an amazing ability to withstand rot. Because of this characteristic, the Romans hollowed out elm wood and used it as water pipes.

Another feature of the elm is that its grain crosses. This makes it very difficult to split as firewood, but it makes it useful in another way. That is that it can be bent without causing it to split. This characteristic has made the elm very useful throughout history in making such things as wagon wheels or long bows. Whereas woods with straight grains split under such bowing, the elm remains whole.

There is a also lot of mythology connected with the elm tree. Even in this verse in Hosea, its only Biblical occurrence, it is used for worshiping idols. Indeed, every tree called by the word *elah* in the Bible seems to be a mark of pagan worship. Almost every occurrence of these trees has some connection to idolatry, just as we find with the elm here in Hosea.

Interestingly, both the American Indians and the ancient Druids also worshipped under elm trees and had similar superstitious beliefs about the elm tree. They both believed that if they burned elm leaves under an elm tree and then drove their cattle through the smoke, it would cure the diseases of the cattle, and also of themselves. Perhaps this is a pagan practice handed down all the way from the time of Hosea, when the people similarly burned incense and worshipped under elm trees.

This practice forms a sharp contrast with the cedar in Ezekiel 17. Under the elm, people attempt to reach up with their smoke to God, but they cannot. By contrast, the true God, pictured by the highest branch of the high cedar on the high mountain in Ezekiel 17, is high and cannot be reached by man. Instead He in mercy reaches down to man. In this way, the elm teaches us the futility of man's effort to reach God. It points us instead to God's mercy to reach down to us, the only way that we can have contact with God.

16
Ezekiel 17 Trees

In chapter 17 of Ezekiel, the prophet is given a riddle from the Lord. *"And the word of the LORD came unto me, saying, Son of man, put forth a riddle, and speak a parable unto the house of Israel." (Ezekiel 17:1-2)* A riddle such as this one is not a mystery. It is not meant to leave us in the dark, nor is it for us to figure out. It is instead given in order to emphasize the importance of what is being said. It is given in parable form. The narrative itself is not necessarily true, but there is a truth that is it teaching. This parable teaches a very important truth using the contrast between the willow and the cedar.

Ezekiel puts forth the first part of the riddle with these words, *"Thus saith the Lord GOD; A great eagle with great wings, longwinged, full of feathers, which had divers colours, came unto Lebanon, and took the highest branch of the cedar." (Ezekiel 17:3)*

In the latter part of this chapter, this riddle is explained. The eagle is Nebuchadnezzar king of Babylon. He is described as *"longwinged."* This pictures the far-reaching spread of Nebuchadnezzar's empire. Just as a longwinged eagle can fly a long distance, Nebuchadnezzar's empire covered most of the known world. This eagle is also described as having feathers of *"divers colours."* This pictures the diversity of Nebuchadnezzar's empire. It was not one consolidated nation of strictly Babylonian culture. Instead it consisted of many different nations, all retaining their own customs and characteristics, but all under the power of Babylon. His was a truly diverse and far-reaching empire, encompassing other nations.

It is interesting that this eagle is also called a *"great eagle."* The largest and greatest eagle known to man is the golden eagle. Again even in this description can be seen a reference to Nebuchadnezzar and Babylon. It was Nebuchadnezzar who was described as a head of gold. After his dream of the great image with the golden head, Daniel had interpreted it to him, saying, *"Thou art this head of gold." (Daniel 2:38)*

This eagle, Nebuchadnezzar, flies in Ezekiel's riddle to Lebanon. Lebanon here is a metaphor for Jerusalem, and the famous cedar of Lebanon is used to picture the inhabitants of Jerusalem. There in Jerusalem Nebuchadnezzar *"took the highest branch of the cedar."* Verse 4 goes on to say, *"He cropped off the top of his young twigs, and carried it into a land of traffick; he set it in a city of merchants." (Ezekiel 17:4)* This highest branch is the king of Judah, Jeconiah, the highest man in Jerusalem. He is taken by Nebuchadnezzar, removed from his high position and carried away into Babylon.

Verse 5 introduces a new character, the seed. *"He took also of the seed of the land, and planted it in a fruitful field; he placed it by great waters, and set it as a willow tree." (Ezekiel 17:5)* This seed is Mattaniah. He is part of the royal seed of Judah in that he is Jeconiah's uncle. Nebuchadnezzar changes his name to Zedekiah and sets him up as a vassal king of Judah in place of Jeconiah.

In this passage he is described as a willow tree. This is a picture of the weakness of his kingship. The willow is weak wooded. It needs a lot of water and won't grow anywhere but close to water. It does not live long. And it is easily broken, but easily replanted. The willow is a transient tree and easily controlled by an authority. For a full description of it, see chapter 52 on the willow tree.

As such a weak and transient tree, the willow is thus a perfect picture of Zedekiah's weak position. He was not intended to be a permanent, sovereign ruler with real power over Judah. But he was set there by Nebuchadnezzar in order to support Babylon. Nebuchadnezzar, in his ruling over many diverse nations, always set up kings over the various nations that he could control and that would support him. Such was the case with Zedekiah. He was chosen to be in charge of Judah in order that he might produce goods that would flow to Babylon.

For a while Zedekiah fulfilled this desire, though perhaps not as fully as Nebuchadnezzar had hoped. *"And it grew, and became a spreading vine of low stature, whose branches turned toward him, and the roots thereof were under him: so it became a vine, and brought forth branches, and shot forth sprigs." (Ezekiel*

17:6) Not even a very good willow tree, Zedekiah became a vine of low stature. He was not a tall, flourishing tree. He produced only some things for Babylon. But to Nebuchadnezzar it was better to have a low and producing willow than a tall cedar that didn't produce anything for his benefit.

Zedekiah at first turned his branches toward Nebuchadnezzar. This he did in faithfulness to the covenant formed between them, that his goods would flow to Babylon. His roots remained planted in Israel, firmly underneath him, but his branches turned toward Babylon.

Long before the captivity, when the Lord established Israel as a nation, Israel's substance flowed to the Lord. Their branches were turned toward the Lord. But when Israel rebelled, the Lord sent Babylon to take them captive. Now in their captivity they had made a covenant with Babylon, and because of their rebellion, their substance was required to flow to their captor, Babylon.

Yet even in this judgment, the Lord gave Israel a second opportunity to repent, as He so often does for His people. Israel had made a covenant with Babylon, to direct their substance there, to shoot their branches in that direction. Here under Zedekiah, the Lord offered them a second chance, returning them to the land and giving them opportunity to obey Him by keeping this covenant.

If they had only taken this opportunity, the Lord both could and would have restored Israel as a proper nation eventually. But just as we often do, Zedekiah and the children of Israel failed to use this opportunity. They did not obey the Lord and keep their covenant when they were offered a second chance.

In verse 7 a new eagle appears. *"There was also another great eagle with great wings and many feathers: and, behold, this vine did bend her roots toward him, and shot forth her branches toward him, that he might water it by the furrows of her plantation. It was planted in a good soil by great waters, that it might bring forth branches, and that it might bear fruit, that it might be a goodly vine." (Ezekiel 17:7-8)*

This eagle was much like the first, Babylon. But it lacked the long wings. Its kingdom was not as far-reaching as Babylon's.

This eagle was Egypt, a large kingdom, but not as large as Babylon.

Zedekiah now made a league with Egypt, hoping that Egypt would help him break his tie with Babylon. Zedekiah may have thought this was a wise political decision. But it is never right to use the world's means in order to circumvent the Lord's will. The Lord's way was for Zedekiah to keep his covenant, and then in the Lord's time He would deliver Israel from Babylonian rule. But Zedekiah chose his own craftiness above God's commandment.

Toward Egypt Zedekiah turned, not only his branches, but also his roots. *"This vine did bend her roots toward him, and shot forth her branches toward him." (Ezekiel 17:7)* Zedekiah gave Egypt his whole heart, not just his outward substance. We notice also that the roots are underground, unseen to the eye of man. This was an underhanded league, made secretly between Zedekiah and Egypt. Secretly Zedekiah sent his roots and his ambassadors to Egypt to help him out of his situation.

Applied personally, this picture of roots and branches gives us a clear picture of how we are to direct our whole lives toward the Lord. If we turn toward Christ with our whole body, which we can see, and with our whole soul, which we cannot see, Christ will prosper us.

But Zedekiah failed to do so. In his underhanded league, he may have thought he was wise, but he was in direct violation of the Lord's clear commandment. God commanded that Israel serve Babylon and not resist. This was the key to their establishment again in their land. *"But the nations that bring their neck under the yoke of the king of Babylon, and serve him, those will I let remain still in their own land, saith the LORD; and they shall till it, and dwell therein." (Jeremiah 27:11)*

Violating this command and relying upon his own human wisdom, Zedekiah will be taken in his own craftiness. *"For the wisdom of this world is foolishness with God. For it is written, He taketh the wise in their own craftiness." (I Corinthians 3:19)* Zedekiah is now in league with two nations. He is burning the candle at both ends, and he must inevitably be burnt.

For keeping his covenant, Zedekiah was in an ideal position. *"It was planted in a good soil by great waters, that it might bring forth branches, and that it might bear fruit, that it might be a goodly vine." (Ezekiel 17:8)* He had been placed by Nebuchadnezzar in a good land, in order that he might bring forth fruit both for Babylon and Israel. Nebuchadnezzar would have left him alone to prosper had he only kept his covenant and rendered his substance to him. But Zedekiah didn't leave well enough alone. He thought he knew better and was wiser.

In verse 9 a question is asked. *"Say thou, Thus saith the Lord GOD; Shall it prosper? shall he not pull up the roots thereof, and cut off the fruit thereof, that it wither?" (Ezekiel 17:9)* With the good land and situation, would the crafty king and his kingdom prosper?

The immediate answer from the Lord is a resounding no and a pronouncement of destruction. *"It shall wither in all the leaves of her spring, even without great power or many people to pluck it up by the roots thereof." (Ezekiel 17:9)* Even the roots will be plucked up. No, he will not prosper, for he did not obey the Lord.

As Christians, we must be careful lest this same judgment come upon us. Christ has planted us here on earth like willow trees to produce fruit for Him. But as willows are weak, temporary, and easily moved, so are we as mortals. Let us not fail Him by not producing fruit. If our lives merely cumber the ground, we may, like Zedekiah, be cut down. Let us instead fulfill the purpose for which we were planted and produce fruit for His glory.

The previous chapter in Luke also gives insight into the judgment upon Zedekiah. *"For unto whomsoever much is given, of him shall be much required: and to whom men have committed much, of him they will ask the more." (Luke 12:48)* Much was committed to Zedekiah. He was set in a good land with the nation of Judah under his rule. Therefore much was required of him. Did he meet this requirement? Would he prosper? The answer is an emphatic no. He would wither and be plucked up by the roots.

The next few verses of Ezekiel 17 are an explanation of this portion of the parable. We pick up the narrative at verse 22. *"Thus saith the Lord GOD; I will also take of the highest branch of the high cedar, and will set it; I will crop off from the top of his young twigs a tender one, and will plant it upon an high mountain and eminent." (Ezekiel 17:22)*

Here the Lord gives a contrast with the first part of the chapter. Nebuchadnezzar had cut off the branch, but the Lord sets up the branch. Nebuchadnezzar had taken the highest branch of a cedar tree. But the Lord Himself takes the ***highest branch*** of the ***high cedar***. In Jerusalem the highest and most prominent cedar was David's kingly line. From this tree the Lord takes the highest branch. This is none other than the Messiah, Jesus Christ.

This cedar the Lord plants *"upon an high mountain and eminent."* This is in stark contrast to the planting of the willow tree. The willow was planted by the water, but this tree is planted in a dry place. In contrast to the willow, the cedar is a hardy tree. It is long-living. It grows slowly and steadily, in contrast to the fast but shallow growth of the willow tree.

The slower a tree grows the stronger it is. This applies to Christian growth as well. When a Christian grows fast right away, he soon wearies and eventually stops growing altogether and perhaps even leaves the church. But when a Christian grows slowly, he step by step increases in the strength he needs. He becomes a strong Christian, able to endure in the long run.

The strong, long-living, evergreen cedar tree, the Messiah, is planted by the Lord in a prominent mountain. At His planting, He is *"a tender one,"* planted in the dry ground of a high mountain. This picture is also used by Isaiah to describe the Messiah. *"For he shall grow up before him as a tender plant, and as a root out of a dry ground." (Isaiah 53:2)*

Ezekiel cannot be speaking merely of the kingdom, but of Christ Himself. Zerubbabel would reestablished the kingdom under the decree of Cyrus king of Persia. This was a partial fulfillment in the restoration of David's line. But it was only temporary, and the throne would again be vacant for centuries. Ezekiel, however, is referring to a permanent kingdom, a slow-growing, long-living cedar tree planted in a prominent mountain as an eternal king. This can be only the kingdom of Jesus Christ Himself.

This ***highest branch*** of the ***highest cedar*** is elevated still higher. *"In the mountain of the height of Israel will I plant it."* *(Ezekiel 17:23)* It is planted in the very height of Israel. Jerusalem is Mount Zion, the high place where the Lord has chosen that His kingdom would be established. One cannot be any higher than this plant. It is the ***highest branch*** of the ***highest cedar*** and is planted on the ***height of Mount Zion***. One cannot get any closer to Heaven.

The rest of this verse goes on to describe the fruit that comes from Messiah. *"And it shall bring forth boughs, and bear fruit, and be a goodly cedar: and under it shall dwell all fowl of every wing; in the shadow of the branches thereof shall they dwell."* *(Ezekiel 17:23)*

The boughs are the apostles, pastors, teachers, and missionaries that come forth from the cedar. The fruits are the souls saved from the ministry of the branches. Under the tree dwell all the fowls of every wing. These are all nations, Jew and Gentile. All nations receive the blessing of Christ, and all will one day be brought under His sovereign rule.

Revelation describes this great day. *"After this I beheld, and, lo, a great multitude, which no man could number, of all nations, and kindreds, and people, and tongues, stood before the throne, and before the Lamb, clothed with white robes, and palms in their hands."* *(Revelation 7:9)* All nations will stand here, all gathered under the cedar tree which God has set up.

In this day, even the rebellious nations will be conquered and subdued under Christ's rule. Ezekiel goes on to prophesy, *"And all the trees of the field shall know that I the LORD have brought down the high tree, have exalted the low tree, have dried up the green tree, and have made the dry tree to flourish: I the LORD have spoken and have done it."* *(Ezekiel 17:24)* The rebellious seem to be high and green trees, but they will be made low and dry. On the other hand, Israel seems to be a low and dry tree, but it will be restored as a nation again and will be made high and green. God is able to do this miracle, to make the dry green and the low high.

In the world today, it seems that Satan is flourishing. He is the prince of the power of the air, and the world seems to be under his sway. Christians do not seem as prominent as they once were. They are dry and low. However they are not dead and will once day prosper again in God's timing.

17
The Fig Tree

The fig tree is one of the most oft-mentioned trees in the Bible. Throughout the Bible, the term "fig" or "fig tree" appears 64 times, and we have already come across several of these references in the earlier chapters of this book. The Hebrew word for the fig tree is *taen* (תאן), and the Greek word is *suke* (συκη).

As we study the fig tree in the Bible, we find that it almost always represents the nation of Israel. This is especially true of the term "fig tree," not necessarily of the term "fig" when it is referred to simply as a fruit, not a tree. For example, when Abigail brought David *"two hundred cakes of figs,"* (I Samuel 25:18) this was probably not representative of Israel, but rather a simple statement of fact. On the other hand, the term "fig tree" does usually represent Israel.

The first mention of the fig tree is very early, in Genesis 3:7. After Adam and Eve sinned, *"the eyes of them both were opened, and they knew that they were naked; and they sewed fig leaves together, and made themselves aprons." (Genesis 3:7)* Aside from the tree of life and the tree of the knowledge of good and evil, the fig tree is the only specific kind of tree named in the garden of Eden. This is significant. It is the only tree that we have today that we know for sure was in the garden of Eden. Whether or not we still have this tree in the same form I don't know, but we do know that we still have at least one of the kinds of trees that were originally in the garden of Eden.

The fig is used for various purposes throughout the Bible. In II Kings, it is used for medicine. In I Samuel and Jeremiah, it is used as a food. It is used symbolically in several places to picture peace and prosperity, with such phrases as *"But they shall sit every man under his vine and under his fig tree; and none shall make them afraid: for the mouth of the LORD of hosts hath spoken it." (Micah 4:4)* The fig is also used often in parables in order to teach spiritual truths.

The fig tree is unusual in that it produces two sets of fruit per year, one early first fruit, and one later summer fruit. Very early in the season, before either leaves or flowers appear to the eye, it produces the first set of fruit, called taqsh, and called in Hebrew *bikkuwrah* (בכורה), or "firstripe fruit." This is the fig tree's sweetest and best fruit. Later, when the leaves have come and the flowers have formed, the summer fruit appears.

The *bikkuwrah*, or first fruit, is a picture of the fruit that the Lord desires from His people Israel, and also from His church, from Christians. It is formed by a flower hidden within a tube. This flower then dissolves into a jelly-like substance, which becomes the sweet fruit. The flower is never seen, but does its work quietly and without show.

In the same way, the Lord wants His people to bear fruit, but He wants it to be done without show and fanfare, without showy flowers and leaves. He doesn't want us to put on bells and to sound horns when we tithe or perform good works. Rather, the sweetest and best fruit is that which comes quietly and without show, like a flower hidden in a tube. The importance is laid upon the fruit, not upon the fruit-bearer.

The fig is noted for having the deepest roots of any tree in the world. One fig tree in Transvaal, South Africa holds the record, with roots that are 393 feet deep.

The banyan tree has vast, spreading roots that can cover several acres, and is also in the *Ficus* family with the fig, as is the mulberry. But this in no way means that they are all the same. They simply share some common characteristics, such as root depth.

The banyan has aerial roots that descend from its branches to the ground. One example is a banyan tree in the Calcutta Botanical Gardens. This tree, in the *Ficus* family and of the type *bengalensis*, covers four acres. The natives of the Himalayas use the banyan tree in an unusual way. They tie several of its roots on one side of a river and several on the other side of the river. As these roots grow, they fuse together and thicken to form a bridge over the river.

It is interesting to note that with all its uses in the Bible, the fig tree is never used to construct idols. The other prominent trees in the Bible are used for this purpose, the cedar, the cypress, the ash, the oak, but never the fig. This is because the fig tree has very weak wood, which rots very easily. Resistance to rot is one thing desired in an idol so that it will last a long time, as we have noted in previous chapters. But the weak and rot-prone wood of the fig is not useful for such a purpose. Instead, its useful part is its fruit, not its wood.

A brief survey of the fruits of these other trees makes this clear. Acorns from the oak, poplar seeds, cedar seeds, and cypress seeds are relatively not useful when compared with the fruit of the fig tree. They, like all fruits, are useful in reproduction of their own kind, but that is about all. But the fig exceeds even such fruit as the olive in usefulness. For unlike the olive, the fig is sweet. The fig fruit is thus very important. It is useful in reproduction, and also in providing both edible and sweet sustenance.

An interesting use of the fig is that found in II Kings 20. Here King Hezekiah was *"sick unto death." (II Kings 20:1)* But after he prayed for his life to be spared, the Lord sent Isaiah the prophet and gave him a miraculous recovery through the fig. *"And Isaiah said, Take a lump of figs. And they took and laid it on the boil, and he recovered." (II Kings 20:7)*

This is miraculous because it is not the natural medicinal use of the fig. Naturally, the fig can be used as a medicine for internal ulcers, but it cannot be used for such external afflictions as boils or carbuncles. But here it is used to heal Hezekiah's boil. Thus we see that this healing is no mere natural remedy, but a true miracle from the Lord. Only He could make the fig effective in this case. It was not the fig that cured Hezekiah, but the Lord.

This is also a picture of God's coming miraculous healing of Jerusalem. In the previous verse, God had given this reason for intervening in this case, *"I will defend this city for mine own sake, and for my servant David's sake." (II Kings 20:6)* God loves Jerusalem, the city of David, though today it has turned its back upon Him. But just as He used figs to heal Hezekiah miraculously, He will one day heal Jerusalem, and the nation of Israel that it represents.

Another interesting Biblical appearance of the fig is found in Jeremiah's vision of Jeremiah 24. *"The LORD shewed me, and, behold, two baskets of figs were set before the temple of the*

LORD, after that Nebuchadrezzar king of Babylon had carried away captive Jeconiah the son of Jehoiakim king of Judah, and the princes of Judah, with the carpenters and smiths, from Jerusalem, and had brought them to Babylon." (Jeremiah 24:1)

In this vision, the Lord used figs to teach Jeremiah the blessings of obedience. Jeconiah, the king of Judah, had obeyed the Lord and submitted to His chastening for sin by going into captivity. But Zedekiah, his brother and the new king of Judah, had remained in the land, rebelling against God's chastening hand. To show the contrast, God showed Jeremiah two very different baskets of figs. *"One basket had very good figs, even like the figs that are first ripe: and the other basket had very naughty figs, which could not be eaten, they were so bad. Then said the LORD unto me, What seest thou, Jeremiah? And I said, Figs; the good figs, very good; and the evil, very evil, that cannot be eaten, they are so evil." (Jeremiah 24:2-3)*

The Lord then told Jeremiah the meaning of these two baskets of figs. *"Again the word of the LORD came unto me, saying, Thus saith the LORD, the God of Israel; Like these good figs, so will I acknowledge them that are carried away captive of Judah, whom I have sent out of this place into the land of the Chaldeans for their good." (Jeremiah 24:4-5)*

The captivity was meant for their good. Jeconiah and those who submitted to it, repenting of their sins, would be made sweet like the good figs. These figs were the *bikkuwrah*, the first ripe figs, the sweetest and best. Jeremiah said that they were *"very good,"* unusually good. This abundant goodness would be the result of obedience and of submitting to God's loving chastening.

But for Zedekiah and those who remained in Judah, rebelling against God's command, God made a very different pronouncement. *"And as the evil figs, which cannot be eaten, they are so evil; surely thus saith the LORD, So will I give Zedekiah the king of Judah, and his princes, and the residue of Jerusalem, that remain in this land, and them that dwell in the land of Egypt." (Jeremiah 24:8)* These rebellious ones would be like the evil figs, the figs that Jeremiah called *"very naughty figs, which could not be eaten, they were so bad."* They had rejected the command to go into captivity, they had refused God's loving hand of correction, and now they must be left to bear the consequences of their own sin.

Micah also shows the superiority of the early fig over the later fig. He laments, *"Woe is me! for I am as when they have gathered the summer fruits, as the grapegleanings of the vintage: there is no cluster to eat: my soul desired the firstripe fruit."* (Micah 7:1) The word translated "firstripe fruit" is *bikkuwrah*, meaning literally the early fig. Micah has the summer fruit of the fig, but it cannot compare with the firstripe fig. He longs for the *bikkuwrah*, the firstripe, the best.

Habakkuk uses the fig to show God's faithfulness even in the bleakest of circumstances. Habakkuk says, *"Although the fig tree shall not blossom, neither shall fruit be in the vines; the labour of the olive shall fail, and the fields shall yield no meat; the flock shall be cut off from the fold, and there shall be no herd in the stalls: Yet I will rejoice in the LORD, I will joy in the God of my salvation."* (Habakkuk 3:17-18)

Because of war and famine, these fig trees in Habakkuk have no blossoms. But this means not only that they bear no fruit this year, but also that they will bear no fruit the next year. Trees need leaves and flowers to store energy in their roots for the next year. But in a hot summer, leaves fall off the trees, and the roots have nothing to give them energy to store for the next year. So the next year is a bad year for fruit, even if the weather is good by that time, because the roots have no energy to give to fruit production. Thus for Habakkuk it will probably take a couple of years to recover from this famine.

The picture is all the more bleak when we consider its symbolism. The fig is the symbol of sweetness and the olive of gladness and cheerfulness. But both are failing due to the war that has come over the land. This is often the case after war. There may be a brief period of rejoicing over the victory, but on the whole, there isn't much gladness or rejoicing for quite a while.

Yet in spite of all of this, Habakkuk says, *"Yet I will rejoice in the LORD, I will joy in the God of my salvation."* (Habakkuk 3:18) Having just come through a devastating war, with at least two years of famine still ahead, Habakkuk still rejoices in the Lord. He knows that no matter what the circumstances are, he can trust the Lord to sustain him and bring him through the trouble, for it will not last forever. In this he rejoices.

Abram, Jacob, and Elimelech all had to learn this lesson the hard way. They left the land of God's choosing in time of famine. They did not trust God to sustain them when the circumstances looked bleak. As a result, they were judged for this unbelief and disobedience. Because they left God's will, God removed His sustaining hand that would have brought them through the famine, and they suffered lack and judgment. Had they stayed in God's will, in the place where God had put them, they would have found what Habakkuk found and what we will find if we will likewise trust God in spite of the circumstances, that God will sustain us if we are in His will.

The fig tree is used in Joel to show God's judgment upon the land of Israel for her sin. Joel says, *"For a nation is come up upon my land, strong, and without number, whose teeth are the teeth of a lion, and he hath the cheek teeth of a great lion. He hath laid my vine waste, and barked my fig tree: he hath made it clean bare, and cast it away; the branches thereof are made white." (Joel 1:6-7)*

The final mention of the fig is in the final book of the Bible, Revelation, just as its first mention was in the first book, Genesis. It spans the entire Bible, from the first book to the last. Revelation 6:13 says at the opening of the sixth seal, *"And the stars of heaven fell unto the earth, even as a fig tree casteth her untimely figs, when she is shaken of a mighty wind." (Revelation 6:13)*

Untimely figs are figs that are not ripe. We have all no doubt seen ripe apples under an apple tree in the fall. Fruits fall to the ground when they become ripe. The attachment between them and the branch become less and less until they finally break off and fall. But unripe fruit has a very strong attachment to the branch. For it to fall, it takes a very mighty wind. The wind spoken of in Revelation is no gentle breeze, but wind of hurricane force or even stronger. It would take such to throw *"untimely figs"* from a tree.

This mighty wind is a picture of God's power to cast the stars from heaven. This could refer to literal stars or it could be symbolic of the rulers and powerful men of the earth. Either way, it takes great power to cast them down, just as it takes great power to cast untimely figs from a tree. But God is all powerful. He holds this power and will exercise it at His appointed time.

Before we leave the fig, we take up one last topic that relates to Christ's second coming. This is Christ's parable of the fig in the Olivet Discourse, which is recorded in three Gospels, Matthew, Mark, and Luke.

To lead up to this, we must look at Christ's cursing of the fig tree, and see how it bears upon the second coming. To set the account in its context, Christ had just made His triumphal entry into Jerusalem. The first day in Jerusalem, He did not take immediate action against the greedy moneychangers. *"And Jesus entered into Jerusalem, and into the temple: and when he had looked round about upon all things, and now the eventide was come, he went out unto Bethany with the twelve." (Mark 11:11)* Christ simply observed, *"looked round about upon all things."* Then He departed to Bethany for the night.

"And on the morrow, when they were come from Bethany, he was hungry." (Mark 11:12) Here, returning to Jerusalem from Bethany, Christ saw the fig tree. *"And seeing a fig tree afar off having leaves, he came, if haply he might find any thing thereon: and when he came to it, he found nothing but leaves; for the time of figs was not yet." (Mark 11:13)* Seeing this fig tree without fruit, Christ then cursed it. *"And Jesus answered and said unto it, No man eat fruit of thee hereafter for ever. And his disciples heard it." (Mark 11:14)*

Why did Christ curse this tree that had no fruit? Verse 13 says that *"the time of figs was not yet."* Why curse the tree if it is not the time for figs? An understanding of the timetable of this event and of the nature of the fig tree reveals the reason behind this curse.

This cursing of the fig tree took place just a few days before Christ's crucifixion. Since the crucifixion occurred at Passover, this event would have taken place some time in March or April. This accords with the fact that it was not yet time for figs. But at this time of year, before the tree's full time for figs, even before or while it puts out leaves, the fig tree also puts out its first small fruit. This is the first fruit, the *bikkuwrah*, that we mentioned earlier. It is an almond-shaped knob and is also called a taqsh. It is from this fruit that the true fig comes about six weeks later.

In Biblical times, this taqsh was available to be eaten by the hungry or poor. Of course, they could not all be eaten, or none would have an opportunity to mature, and there would be no true figs at the time of figs. But the few poor and hungry peasants that passed by could eat of these small, almond-like fruits.

Nahum also referred to this early fruit of the fig tree. *"All thy strong holds shall be like fig trees with the firstripe figs: if they be shaken, they shall even fall into the mouth of the eater."* *(Nahum 3:12)* This firstripe fig is the taqsh, the first and earliest fruit to come on the tree.

It was this fruit that Christ sought on the fig tree. He knew that if He could find no taqsh on the tree at this time of year, then at the time of figs, it would not have any figs either. Therefore, He cursed the tree, knowing that it would not bear fruit.

The timing of the events in Matthew and Mark seem contradictory. In Matthew, it seems that Christ entered Jerusalem and cleansed the temple on the same day, and then cursed the fig tree the next day and it withered immediately. In Mark, it seems that he entered Jerusalem, cursed the fig tree the next morning, cleansed the temple that day, and the fig tree didn't wither until the day after.

Yet this is only a seeming contradiction, not a real one. Matthew's account is a condensed version, focusing on the place, not the time. He tells of Christ's triumphal entry and cleansing of the temple in the same passage because they occurred at the same place. He also tells of the cursing of the fig tree and its withering in one passage because they occurred at the same place. Mark goes into more detail and gives attention to the exact chronology and timing of these occurrences.

Knowing this timing and the surrounding events is important for understanding Christ's action in cursing this fig tree. He was not angrily out of control. He did not "fly off the handle." This was rather a carefully measured action.

Had Christ been out of control, He would have immediately overthrown the tables upon finding the moneychangers there.

Instead He first, *"looked round about upon all things."* (Mark 11:11) He waited that entire day, and it wasn't until the next day that He took action, not in rage, but in carefully considered justice.

"And they come to Jerusalem: and Jesus went into the temple, and began to cast out them that sold and bought in the temple, and overthrew the tables of the moneychangers, and the seats of them that sold doves; And would not suffer that any man should carry any vessel through the temple." (Mark 11:15-16) This was not an uncontrolled outburst of passion. Christ went on to explain calmly His reason for this righteous vengeance. *"And he taught, saying unto them, Is it not written, My house shall be called of all nations the house of prayer? but ye have made it a den of thieves."* (Mark 11:17)

For this action, the Pharisees then became uncontrollably angry at Him. *"And the scribes and chief priests heard it, and sought how they might destroy him: for they feared him, because all the people was astonished at his doctrine."* (Mark 11:18) Unruffled and in full control, Christ then departed from the scene. *"And when even was come, he went out of the city."* (Mark 11:19)

Having shown Christ's controlled action in cleansing the temple, Mark then picks up the story of the fig tree once again. *"And in the morning, as they passed by, they saw the fig tree dried up from the roots. And Peter calling to remembrance saith unto him, Master, behold, the fig tree which thou cursedst is withered away."* (Mark 11:20-21)

The cursing of the fig tree, like the cleansing of the temple, is not a mere outburst of anger. It is an important lesson, woven into the account of the temple cleansing and explained by it. Throughout the Bible, the fig tree is almost always used as a picture of the nation of Israel. Here, when Christ entered Jerusalem, He came to Israel expecting to find fruit. But He found only outward pomp. He found self-righteous priests and Pharisees. Outwardly they looked good and holy, but they were interested only in making money. They had outward, showy leaves, but they had no inward fruit.

There is much similarity in purpose between this account of the cursed fig tree and Luke's parable of the unfruitful fig tree that we will look at in chapter 19 on fruit. Both picture Christ seeking fruit on the nation of Israel, but finding none. Yet they are distinct and different. Luke's is a parable. This one is a true account of a real fig tree.

An interesting fact about this tree is that, after Christ cursed it, it was withered by the roots. *"And in the morning, as they passed by, they saw the fig tree dried up from the roots." (Mark 11:20)* This is very unusual. Usually it is the upper, visible part of the tree that withers and dries up. But here, the roots, which none can see, withered up.

Again, this is a picture of the withering of Israel. Outwardly Israel looked prosperous and holy. But in the roots, where none could see, there was dryness and withering.

Israel was intended to be the root and power of the church. It was through Israel that Jesus, the head of the church, came to earth. It was to Israel that He first revealed Himself. But the religious leaders of Israel were all outward show. Thus Christ withered up this root. The power and strength of Israel to nourish the church was taken away.

This cursing of the fig tree is the first record of Christ performing a miracle to destroy or curse something while in human form. Usually while He walked on earth, He was healing and helping and restoring. Why then did He curse this fig tree? Did He hate trees? No, He cursed this tree for a specific reason. He could have commanded the tree to produce fruit. But instead He condemned it to eternal barrenness. In this way, He pressed indelibly and unforgettably upon the hearts of His disciples, and of us, the important lesson that we must bear fruit.

A question that often arises about this fig tree is, was the tree dead once it withered? The root of the word "withered" in this passage is the Greek word *xeros* (ξηρος), which means simply "lack of water."

The Latin word *xeros*, a cognate of the Greek, is used today in the term "xeroscaping." Xeroscaping is landscaping that uses plants that need very little water, such as cactus plants. It is done mostly in the southwestern United States. One of the plants used in xeroscaping is the creosote plant. It needs very little outside watering because it collects even the smallest traces of water in the soil. In fact, this plant is so adept at taking water out of the ground that no plants can grow around it. It leaves the ground completely dry and unable to sustain any other plants.

Thus, based upon the word *xeros*, when Christ caused the fig tree to wither, He was simply depriving it of water. This is different from death. Death is final separation, the changing from one form to another. If the fig tree had died, the Bible would have said that it died. Instead it says that it withered. This fig tree was not dead, but only withered and dry.

So the fig tree, Israel, was dead in the sense that as a whole it was separated from Christ, but it still had living sap within. A live tree can become so dry that by all outward appearances, it looks dead. This was the case with Israel. It was withered and dry, cut off from the living water of Christ, yet there was a remnant of true believers within. This remnant of Israel still exists today. Israel is not dead and will not be entirely destroyed. The Lord will save the believing remnant and restore them as His people.

We read an interesting phrase about this tree in Mark's account, that *"the time of figs was not yet." (Mark 11:13)* This phrase gives great insight into the symbolism of this passage. While the tree should have had the firstripe figs, it was not yet the time of the later, summer figs. In the same way, it was not yet the time for Christ to accomplish His triumphant redemption of Israel. This was only His first coming, the time that He must suffer and die for man's sins. Israel was looking for the later fruits, for a reigning Messiah. This is why they spread palm branches in His way, the symbol of victory and triumph. They thought that Christ would triumph over Rome and over their oppressors. But it was not time for this yet. *"The time of figs was not yet,"* and the time of full redemption was not yet.

However, even though it was not the time of full redemption, Israel was still supposed to produce fruit. She was to produce the first fruits. The same is still true for us today. Though the time of figs is still not yet, and we still await Christ's second coming, we must produce the first fruits in the meantime. We must remain steadfast, telling others about Christ, and thereby bearing fruit. Christ will produce the full fruit in His Own special time, but we must still abide in Him and produce fruit until that time.

Another interesting phrase is found in Christ's curse itself in Matthew, *"Let no fruit grow on thee henceforward for ever."* *(Matthew 21:19)* The phrase *"henceforward for ever"* is in the Greek *eis ton aiona* (εις τον αιωνα), and means literally "until the age" or "until the time." Applying this to Israel, Christ has not cut them off forever for their unfruitfulness in the first fruits, but He has placed a curse upon them until the time of His second coming, when they will again be permitted and expected to bear the later fruits.

To sum up the parable, Christ came at His first coming, looking for fruit on Israel. But finding none, He cursed Israel to wither, but not to die, causing them to wither only "until the time" of His second coming. At this second coming of Christ, Israel will be given a second chance to bear fruit, as Christ comes seeking the later fruits.

This brings us to Christ's parable of the fig in the Olivet Discourse. Matthew gives the parable in these words, *"Now learn a parable of the fig tree; When his branch is yet tender, and putteth forth leaves, ye know that summer is nigh: So likewise ye, when ye shall see all these things, know that it is near, even at the doors."* *(Matthew 24:32-33)*

Mark gives a very similar reading, *"Now learn a parable of the fig tree; When her branch is yet tender, and putteth forth leaves, ye know that summer is near: So ye in like manner, when ye shall see these things come to pass, know that it is nigh, even at the doors."* *(Mark 13:28-29)*

Luke's wording is slightly different, but still similar, *"And he spake to them a parable; Behold the fig tree, and all the trees; When they now shoot forth, ye see and know of your own selves that summer is now nigh at hand. So likewise ye, when ye see these things come to pass, know ye that the kingdom of God is nigh at hand."* *(Luke 21:29-31)*

There are two possible interpretations of the parable of the fig tree. The first one draws upon the verses immediately prior to the parable in Matthew and Mark. Matthew says, *"And he shall send his angels with a great sound of a trumpet, and they shall gather together his elect from the four winds, from one end of heaven to the other." (Matthew 24:31)* And Mark says similarly, *"And then shall he send his angels, and shall gather together his elect from the four winds, from the uttermost part of the earth to the uttermost part of heaven." (Mark 13:27)*

This interpretation is that these verses refer to the rapture, the picking of the first fruits just before Christ's second coming. If this were the case, the first fruits would picture those already saved, and the later fruits would picture those that will be saved during the tribulation. The fig, having both first fruits and leaves at the same time, would in this interpretation both picture the rapture and still fulfill its role in the parable, by putting forth leaves that indicate Christ's second coming is near.

The parable in Luke also could fit this interpretation. In this Gospel, the parable is not limited to the fig tree, which pictures Israel, but it is broadened with these words, *"Behold the fig tree, and all the trees." (Luke 21:29)* Instead of limiting the application to Israel, Luke's Gospel applies the parable to all nations. Christians in all nations are raptured out, and all nations must be ready for Christ's second coming.

While this is a possible interpretation, I doubt that it is the true one. I tend to lean toward the second interpretation, which covers a broader scope of history. This interpretation leans on the fact that the fig tree often, though not always, represents Israel. It interprets this parable in light of Christ's cursing of the fig tree, which had just happened a day or two before this parable was given. At Christ's first coming, He found no first fruits on Israel, just as He found no first fruits on the fig tree He cursed. At His second coming, when the leaves are put forth, He will be looking again for fruit on Israel, the later, summer fruits. The question is left hanging, "Will He find the summer fruits at His second coming?"

Using Luke's parable of *"all the trees"* again, this penetrating question is applied, not just to Israel, but to all nations. This makes the question even more personal, "Will Christ find summer fruits on us when He comes?"

18
The Fir Tree

Commentaries differ widely on the identity of the fir tree. Some say that it is a pine, some say a cypress, some a cedar, and some a juniper. But there are problems with all of these.

The pine is already mentioned elsewhere in the Bible, and its needles and cones are different from those of the fir. Whereas fir needles are flat, pine needles are round. Pine needles also grow in bundles of two, three, and five, never having one needle growing by itself like the fir. For the cones, fir cones stand up, but most pine cones hang down.

Some argue that the fir is the pine because both are in the *Pinaceae* family. This is true, but we must be very clear when we say that trees are in the same family. This is not at all meant to imply that they are the same in their properties. For example, parsley and hemlock are in the same family. But while hemlock is poisonous, parsley is well-known to be edible and even nourishing. To say that trees are in the same family is to say little.

The cypress is not a viable option either. It is in an entirely different family.

The cedar also fails the test in that it bears no cones, though it looks similar to the fir. In fact, the fir is related to the *Cedrus* family, but again this does not necessarily mean that they are the same, for many trees within the same family are different. Christians within a church are also part of the same family, but this does not mean at all that they are all the same as one another. They retain individual differences though they are in the same family. The cedar is also found throughout the Bible. If it were the same as the fir tree, why would it appear in the Bible under two different names?

Finally, the juniper, though an evergreen, doesn't have true needles. Instead, it has scale-like leaves that look more like pin feathers. And it also is in a different family than the fir.

Not only are the English words for each of these trees different, but also the Hebrew words. Even in the original language, God makes a distinction between these kinds of trees.

The Hebrew word for fir is *barowsh* (ברוש), and it is always translated fir. Once again, I think the 1611 King James translators got this word right, and that it is a true fir, the fir we classify with the Latin term *Abies*. When the Bible says fir, it means fir.

The fir was used for many things in the Bible as we have already found in our study. It was used for musical instruments in II Samuel. *"And David and all the house of Israel played before the LORD on all manner of instruments made of fir wood, even on harps, and on psalteries, and on timbrels, and on cornets, and on cymbals." (II Samuel 6:5)* Fir wood was used for the floors, walls, and doors of the temple, as recorded in I Kings and II Chronicles. It was used for rafters and beams in Song of Solomon. And it was used for ship boards in Ezekiel. Fir wood is ideal for this kind of construction because it is a strong wood, but has enough flexibility to give a little.

Its only drawback in construction is its susceptibility to rot when it is exposed to the elements and to fluctuation between being wet and being dry. This susceptibility to rot makes it less desirable for outdoor construction, which is why fir wood was used mainly for indoor framing, such as rafters, walls, floors, and also for musical instruments. Also for this reason, fir wood was not used to make idols.

Still, however, it was used for outdoor ship boards by the Tyrians. Ezekiel records, *"They have made all thy ship boards of fir trees of Senir: they have taken cedars from Lebanon to make masts for thee." (Ezekiel 27:5)* Though ship boards would have been exposed to a lot of getting wet and then drying out, which causes rot, still fir wood was chosen for these boards.

This is because of two things, the durability of fir wood and the ease of replacing deck boards. Fir wood, though it rots relatively easy, is strong and durable for construction. But when it does rot, a ship deck is a good place to have it because it these boards are easily replaced.

It is instructive to note the place from whence the Tyrians took the firs for their ship boards. Some argue that these cannot be true firs because true firs don't grow in the hot climate of Israel. This is true, but the Bible doesn't say the firs came from the hot areas of Israel. These were *"fir trees of Senir." (Ezekiel*

27:5) Senir is a peak in the mountain range of Mt. Hermon, which stretches between Lebanon and the inheritance of the tribe of Manasseh. Often what is called a mountain in the Bible, such as Mt. Hermon, can be both a specific peak and also a mountain range or mountainous area. This particular mountainous area was called Mt. Hermon by the Hebrews and Senir by the Amorites.

This area was high in elevation, reaching nine thousand feet in some places, and therefore was not warm, but often covered with snow. In fact, not only were fir trees brought down from here, but also snow was brought down by the Tyrians in the summer time to cool things off. This mountainous location is the very place where we would expect to find the fir, for it grows best at high elevations and in cool temperatures.

A certain type of fir, the white fir or *Abies concolor*, can occasionally be found in warmer climates, in zones five to eight. But the trees are small, only about thirty feet tall, and it is rare for them to reach even one foot in diameter. But in the north, fir trees reach fifty or sixty feet in height and are much thicker than in the warm south. Fir trees also grow in greater abundance in the cooler temperatures of the north.

Two other passages refer to bringing fir trees from the mountains, and both are very similar to one another. II Kings says, *"By thy messengers thou hast reproached the Lord, and hast said, With the multitude of my chariots I am come up to the height of the mountains, to the sides of Lebanon, and will cut down the tall cedar trees thereof, and the choice fir trees thereof: and I will enter into the lodgings of his borders, and into the forest of his Carmel." (II Kings 19:23)*

Isaiah, with almost exactly the same wording, says, *"By thy servants hast thou reproached the Lord, and hast said, By the multitude of my chariots am I come up to the height of the mountains, to the sides of Lebanon; and I will cut down the tall cedars thereof, and the choice fir trees thereof: and I will enter into the height of his border, and the forest of his Carmel." (Isaiah 37:24)*

Both these passages mention bringing down *"choice fir trees"* from the mountains. This makes sense because firs, as is seen in the alpine fir, flourish best at two thousand to five thousand feet above sea level, the elevation of the lower slopes of the mountains of Senir.

Isaiah mentions the fir as being planted in a very different place. God says, *"I will plant in the wilderness the cedar, the shittah tree, and the myrtle, and the oil tree; I will set in the desert the fir tree, and the pine, and the box tree together."* *(Isaiah 41:19)* Here the fir tree is planted by God *"in the desert."* In fact, all three of the trees planted in the desert, the fir, the pine, and the box, are evergreens. This is a very unnatural place for them to be. But God has the supernatural power to plant these trees there and sustain them.

It is comforting to know that we serve such a God. Though He may place us in the desert, He will not leave us without a respite. He will plant evergreens along the way to refresh us from the dryness and heat. In the desert, it is normally very difficult to find shade. But God plants evergreens like the fir, full of greenness and signs of life, to provide us with rest and refreshment in the deserts of our lives.

The fir is personified as rejoicing earlier in Isaiah. *"The whole earth is at rest, and is quiet: they break forth into singing. Yea, the fir trees rejoice at thee, and the cedars of Lebanon, saying, Since thou art laid down, no feller is come up against us."* *(Isaiah 14:7-8)* This passage speaks of the final restoration of Israel to their land, and of their deliverance from captivity. In this day, oppressors will no longer come against them, as the Romans, Babylonians, and Assyrians had, and they will no more be in slavery to other nations.

Not only do the people rejoice, but even the trees rejoice in this deliverance. No longer will foreign nations come and cut them down to make ships. They are at rest and quiet, and therefore they rejoice, saying, *"No feller is come up against us."* *(Isaiah 14:8)*

In Isaiah 55, the fir is again used to show God's grace in difficult times. *"Instead of the thorn shall come up the fir tree, and instead of the brier shall come up the myrtle tree: and it shall be to the LORD for a name, for an everlasting sign that shall not be cut off."* *(Isaiah 55:13)* In the place of thorns and briers God gives instead firs and myrtles.

The order in which these plants are listed makes an interesting contrast here. Firs are larger than myrtles, just as thorn trees and bushes are larger than brier trees and bushes. Also firs and thorns grow at high elevations in snowy, winter conditions, while myrtles and briers grow in low areas in hot, arid, summer conditions.

This grouping shows God's desire to use different men and women for different areas of His service. God has made different kinds of people for different roles, pastors, teachers, missionaries, evangelists, and lay people, each of whom helps in the Lord's work with his own particular gift. Like the fir and myrtle, some are larger than others and suited for large tasks. Some are smaller and suited for small tasks. Some are suited for high elevations and cold conditions. Some are suited for low areas and hot or dry conditions. Whatever their unique characteristics, God has a specific place and purpose in His service that He has selected each one to fill.

This passage also looks forward to the end of time when the curse of sin will be finally removed. The curse of sin upon the ground in Genesis is what brought thorns and thistles in the first place. But in Revelation, when Christ triumphs over sin, He will remove that curse, and replace the thorns and briers with what is much better, evergreens such as firs and myrtles.

Another place where the fir is mentioned is Isaiah 60. *"The glory of Lebanon shall come unto thee, the fir tree, the pine tree, and the box together, to beautify the place of my sanctuary; and I will make the place of my feet glorious."* (Isaiah 60:13) The three trees named in this verse are the same ones named in Isaiah 41:19. They are all three evergreens. With their seemingly everlasting life, these evergreens picture all true Christians. Every true Christian that has trusted Jesus Christ as Savior has everlasting life. In this sense, Christians are all the same.

But these three trees, though they are all evergreens, are all very different from one another. And so in another sense, Christians are also all very different from one another. Every Christian has different talents and different looks because they all have different purposes for which they were formed. While they are all part of one whole, the Church of Jesus Christ, they each fulfill a unique role within the Church, for which they are uniquely suited.

At Christ's coming, this will continue in heaven on a larger scale. At that day, the individual Christians from many different nations will be gathered into heaven. All that will be there will be Christians, souls who have eternal salvation through the blood of Jesus Christ. But their differences won't vanish in eternity. Different nations will retain their differences and will each have different purposes to fulfill. Different Christians also will still look different. They will be given different crowns and responsibilities, and will fulfill different purposes throughout eternity.

We see this diversity within unity pictured in the nation of Israel as well. Israel is one nation, the nation of Christ's chosen people. But it has twelve different tribes, each with its own task and purpose. For example, Judah has a different role than Levi. Yet both are important, and both are Christ's chosen people.

In Ezekiel 31, the fir is used to teach humility. In this passage, which we will study in chapter 26, Pharaoh had become lifted up with pride. He thought himself great and important. But Ezekiel compared him to the even more powerful king of Assyria. He likened this king to a tree and said, *"The cedars in the garden of God could not hide him: the fir trees were not like his boughs, and the chesnut trees were not like his branches; nor any tree in the garden of God was like unto him in his beauty."* *(Ezekiel 31:8)* The beautiful fir trees could not rival the beauty and splendor of the king of Assyria. Yet God was still more powerful and brought down the king of Assyria from his lofty heights.

Ezekiel used this to teach Pharaoh that he was not as beautiful, powerful, or important as he thought he was. There was another earthly king more powerful than he. Yet even this king was brought low by the yet more powerful and almighty King of Kings. In comparison to Him, Pharaoh was as nothing.

This should serve as a lesson to us as well. We are not as beautiful, powerful, or important as we think we are. There are others more powerful and more beautiful than we. But above all of them is Christ. He is able to bring all of them down. If He can bring down those more powerful than we, what can He do to us?

Hosea's reference to the fir teaches a lesson on fruitfulness. Here God says, *"I will be as the dew unto Israel: he shall grow as the lily, and cast forth his roots as Lebanon. His branches shall spread, and his beauty shall be as the olive tree, and his smell as Lebanon. They that dwell under his shadow shall return; they shall revive as the corn, and grow as the vine: the scent thereof shall be as the wine of Lebanon. Ephraim shall say, What have I to do any more with idols? I have heard him, and observed him: I am like a green fir tree. From me is thy fruit found."* (Hosea 14:5-8)

Notice the future tense of this passage. Such verbs are used as *"will be," "shall grow,"* and *"shall spread."* This is a time of growing, of looking forward to full maturity at some time in the future. The fir tree in verse 8 is *"a green fir tree."* It is still a young and tender fir tree. It is immature, growing, and not yet full-sized. Yet it is fruitful. The Lord says, *"From me is thy fruit found."* (Hosea 14:8)

It is while a tree is young that it is most fruitful in terms of percentage. Though a tree is still supposed to produce fruit when it is older, its best time for fruit production is while it is young.

The same is true of men and women. As we get older, we physically can't do what we used to do. For this reason, we should by this time be spiritually stronger than when we were young, to compensate for our physical weakness. In this spiritual strength, we still should produce all the fruit we are able. But the best years of our lives for producing fruit are the years of our youth. Let us not squander these years, but use them to their fullest by producing abundant and enduring fruit.

Yet we must not forget the true source of our fruit. God says, *"From me is thy fruit found."* (Hosea 14:8) Fruit is not found in our strength. It does not come from us, but only from Christ. We do have the responsibility to draw upon Him, but only He can supply the strength and substance for the fruit. He is the true source, the vine. We are merely the branch drawing upon Him.

Nahum used the fir to describe the magnitude of God's judgment upon Nineveh. Nahum said of Nebuchadnezzar and his

Babylonian army, Nineveh's oppressors, *"The shield of his mighty men is made red, the valiant men are in scarlet: the chariots shall be with flaming torches in the day of his preparation, and the fir trees shall be terribly shaken." (Nahum 2:3)*

The chariots of Babylon were so numerous and powerful that they literally shook the tall and sturdy fir trees about them. This is not figurative language, but literal shaking. We have all no doubt felt the ground shake when a train passes nearby. The same principle was true here on a larger scale. It takes a lot to shake a fir tree. But that is what the chariots of Babylon did. Their noise and vibration were so great that they shook the fir trees, mighty though they were.

Lest we think that Babylon got off easily, we must hasten to say that its turn to shake was coming. Not many years later, God pronounced judgment on Babylon by writing with his finger on the wall at Belshazzar's feast. At this, Belshazzar shook in terror. The Bible records of him, *"Then the king's countenance was changed, and his thoughts troubled him, so that the joints of his loins were loosed, and his knees smote one against another." (Daniel 5:6)* God's judgment is sure and powerful, mighty enough to shake the strongest tree and the greatest monarch.

Zechariah also mentions the fir tree. He says, *"Open thy doors, O Lebanon, that the fire may devour thy cedars. Howl, fir tree; for the cedar is fallen; because the mighty are spoiled: howl, O ye oaks of Bashan; for the forest of the vintage is come down." (Zechariah 11:1-2)*

The fir and the cedar are found together here, as they so often are in the Bible. Both are evergreens, and both are used to represent Israel. But even with their similarities, the cedar still stands out above the fir as more important. It holds a high place of importance in Scripture in that it is used more often than even the fig to represent God's chosen people Israel, especially the kings of Israel.

This comparison is important in shedding light on the meaning of this passage. The fir, the less important tree, is told to howl because the cedar, the more important tree, is fallen. The application is this. If the mighty and more important fall, how much more shall the weaker or less important fall? They have no chance of standing at all. This is accentuated by the fact that the fir tree is more combustible than the cedar tree. If the fire can devour the resistant cedar tree, how much more will it devour the combustible fir?

This passage in Zechariah is prophesying of Israel's rejection of Christ. In that rejection, the leaders of Israel, the important men, would fall. They would lead the way in rejecting Christ. Then the common people, the less important ones, would merely follow their leaders and fall likewise into the same sin of rejection. For this reason, the fir trees, the less important, are told to howl.

One of the most unusual, but also encouraging, references to the fir tree is found in the Psalms. Psalm 104 says, *"The trees of the LORD are full of sap; the cedars of Lebanon, which he hath planted; Where the birds make their nests: as for the stork, the fir trees are her house." (Psalm 104:16-17)*

The stork is a very large bird. A small nest is not a sufficient term for the dwelling place it needs. Rather it needs a large house. When we read of these Biblical trees in this passage, we must remember that these are large trees. The Biblical cedar is not the same as our North American cedar. The cedar of the Bible is a *cedrus*, while that of the United States is a *juniperus*. The cedars and firs of the Bible are much larger trees than those with which we are familiar. These large firs are large enough to form a house for the stork.

But the main thing to notice about this passage is who plants these trees. They are called *"the trees of the LORD,"* and trees *"which he hath planted."* When God plants trees, they are always *"full of sap."* He is the master of arboriculture. His trees never die, and He knows exactly how to care for them. This cannot be said of men. Even men with much experience in growing trees do not hold the key to successful planting. Their plants can and do die at times. Only God's trees remain continually alive and *"full of sap."*

Thus concludes our study of the fir tree. May it be that each one of us is a tree planted by the Lord, a tree full of sap and capable of being used in His service, whether as a nest for the small or as a house for the great.

19
Fruit

We now turn to look at the fruits of the Bible. Many specific fruit trees are mentioned throughout the Bible, the pomegranate, the palm, the fig, and many others. But in this chapter, instead of focusing on specific fruits, we will look at fruit in general throughout the Bible.

Beginning a study like this, often the first thing that comes to mind is the fruit of the Spirit. The fruit of the Spirit is given in Galatians chapter 5. *"But the fruit of the Spirit is love, joy, peace, longsuffering, gentleness, goodness, faith, Meekness, temperance: against such there is no law." (Galatians 5:22-23)*

Jesus manifested all of these qualities in His life. He was the perfect tree and bore perfect fruit. As Christians, we are to be like Jesus Christ, and therefore we also should manifest all of these. But do we? Of course, we must admit that we do not.

That is the reason for the purging of John 15. It is so that we may produce this type of fruit in our lives. As John 15 says, it is so that we may bring forth *"more fruit,"* and eventually *"much fruit."* Of course none of us has all these fruits perfectly yet. But we know Someone Who does. He is our example. It is by Him that we are to measure our fruit.

Ezekiel also speaks of fruit and its application to the Christian life. *"And the first of all the firstfruits of all things, and every oblation of all, of every sort of your oblations, shall be the priest's: ye shall also give unto the priest the first of your dough, that he may cause the blessing to rest in thine house." (Ezekiel 44:30)* The fruit spoken of here is not only the firstfruits, but the first of the firstfruits. It is the best of the best.

The priest that receives the firstfruits is representative of Christ. There is a principle here that God deserves the very best of everything we have. I believe God honors it when we study His Word the first thing in the day. I have found that first thing in the morning is the best time for me to study my Bible. At night I am tired, but in the morning, my mind is fresh, and what I read helps me throughout the day.

We are also to worship the Lord on the first day of the week. And it is the firstborn that receives the inheritance. All through the Bible, we see this principle of God's special blessing upon the firstfruits, and the need to dedicate those firstfruits to Him.

The second part of the verse says to give *"the first of your dough."* We know that this is referring to bread dough, but it is no accident that the word "dough" is used in modern English slang to mean money. Money is to our economy as bread was to the Israelites. It is the basis of daily exchange and the sustainer of life. The first and best of even this necessity must be given to the Lord.

Abraham gave the first of his wealth when he paid tithes to Melchizedek. And the Lord commands the same to all in Malachi. *"Bring ye all the tithes into the storehouse."* *(Malachi 3:10)*

Christ deserves our money and our offerings. It is He Who has given us the health to work and earn money. It is He Who has given us the very job at which we work. It is He Who gives us all that we have, and He deserves the first of His harvest.

In fruit production, it is always the firstfruits that are the best. Paul echoes this principle in II Timothy, *"The husbandman that laboureth must be first partaker of the fruits."* *(II Timothy 2:6)* The first fruits harvested in the season are the most nutritious. As the season goes on, disease and insects begin to harm the later fruits.

Even the fall chrysanthemums illustrate this principle. At first the flowers are full and pretty. But as the season goes on, there are still flowers, but they are not as pretty as the first flowers.

Another fruit spoken of in the Bible is the fruit of lies. This is spoken of by the prophet Hosea. *"Ye have plowed wickedness, ye have reaped iniquity; ye have eaten the fruit of lies: because thou didst trust in thy way, in the multitude of thy mighty men."* *(Hosea 10:13)*

Lying is trying to do something that makes us look good. The fruit of the honeysuckle, which pictured the evil fruit of Matthew 7, is a good picture of this fruit of lies as well. The red berry looks good, but its looks deceive, for it is poisonous and harmful.

The persimmon in the early fall can also picture the fruit of lies. A persimmon at that time of year looks edible, but it is very bitter. The sweetness doesn't come until later.

Just as the persimmon sweetens over time, we as Christians should sweeten as we grow older and mature. While we are babes, we may in our ignorance manifest the bitter fruit of lies. But this should not remain the case. The older we become as Christians the sweeter we should become.

Other fruits are not only bitter or able to cause sickness, but able to cause death. While most of these are not in North America, just a little bite of some of these fruits is enough to kill a person. In the same way, a lie looks good. But if we look further into it, we realize that it is a lie.

Matthew 7 speaks of good fruit versus evil fruit. Those who truly belong to Christ bring forth good fruit. Verse 20 teaches us that we can know people's heart by their fruit. *"Wherefore by their fruits ye shall know them." (Matthew 7:20)*

We have all seen this before. We meet someone and immediately know that they do not have the right kind of fruit. They reveal it in the way they talk, the way they act, and the way they look. Some may object that we can see only the outside, and cannot know what is on the inside. But this verse teaches us that what is on the inside comes out eventually. If a person continually talks, acts, or looks a certain way, if there is no change on the outside, then it reveals that there has been no change of heart on the inside.

It is the inside that determines the outside. Thieves are not thieves because they steal, but they steal because they are thieves. We do what we are. Outward actions are the visible evidence of what is in our hearts. Thus we can easily tell by a person's fruit what kind of tree he is.

Christ tells another parable about fruit in Luke 13. *"He spake also this parable; A certain man had a fig tree planted in his vineyard; and he came and sought fruit thereon, and found none. Then said he unto the dresser of his vineyard, Behold, these three years I come seeking fruit on this fig tree, and find none: cut it down; why cumbereth it the ground? And he answering said unto him, Lord, let it alone this year also, till I shall dig about it, and dung it: And if it bear fruit, well: and if not, then after that thou shalt cut it down." (Luke 13:6-9)*

This parable pictures the nation of Israel. The Lord sought fruit long on the nation of Israel. He sought fruit first at Mount Sinai. When He found none and found them worshipping a golden calf instead, He determined to destroy them. He said to Moses, *"Now therefore let me alone, that my wrath may wax hot against them, and that I may consume them: and I will make of thee a great nation." (Exodus 32:10)* But, like the dresser of the vineyard in this parable, Moses pled that the tree be spared, *"Yet now, if thou wilt forgive their sin--; and if not, blot me, I pray thee, out of thy book which thou hast written." (Exodus 32:32)*

The Lord hearkened to Moses' pleading, and led His people to the border of Canaan. Here He sought the fruit of faith to enter the land. But instead He found unbelief and a refusal to enter. Again, He determined their destruction, saying to Moses, *"I will smite them with the pestilence, and disinherit them, and will make of thee a greater nation and mightier than they." (Numbers 14:12)* But again, Moses interceded and begged that the Lord not yet cut down the tree, *"Pardon, I beseech thee, the iniquity of this people." (Numbers 14:19)* Again the Lord spared the tree.

Many years later when Christ came to earth, He was still seeking fruit on the tree of Israel. For the first three years of His ministry, He labored among the Jewish people, seeking the fruit of faith in Him as Messiah. But still they manifested no fruit. They rejected Him, and thus He was compelled to reject them. He said at the end of this chapter, after the parable, that He would have gathered them under His wings, but *"ye would not!" (Luke 13:34)* Still for almost another year, Christ ministered, sparing the tree, digging about it, dunging it, giving His people opportunity to repent. But their continued refusal to do so eventually led to their being cut off as a whole.

A similar picture of unfruitful Israel is found in Isaiah. *"For the vineyard of the LORD of hosts is the house of Israel, and the men of Judah his pleasant plant: and he looked for judgment, but behold oppression; for righteousness, but behold a cry." (Isaiah*

5:7) When the Lord sought fruit on His people, He found instead oppression and a cry.

What causes this failure to produce fruit? The failure cannot be on the part of the vinedresser. Christ is the perfect vinedresser and the perfect intercessor for sparing His people. The failure is entirely on the part of us as people. It is our responsibility to produce fruit. With the perfect vinedresser and intercessor, we have no excuse not to. Thus when we fail, we must bear the full responsibility and suffer the tragic consequence of rejection.

We look now at some of the things that a plant needs in order to produce fruit. The first is **purging**, as seen in John 15. The second is **water**. This need is found in Psalm 1:3 and Jeremiah 17:8. Plants also must be **weeded** in order to produce fruit, as Mark 4:7,18-19 illustrates. **Fertilizer** also stimulates fruit production as demonstrated in Luke 13:8. And finally in Isaiah 5:1-7 plants need proper **protection** in order to produce fruit.

We will cover the need for **purging** in chapter 39 when we look at pruning in John 15. We now turn our attention to the need for **water**. Water is vital to plant growth and fruit production. Jeremiah 17:8 gives a vivid picture of how the man who trusts in the Lord is like a well-watered plant. *"For he shall be as a tree planted by the waters, and that spreadeth out her roots by the river, and shall not see when heat cometh, but her leaf shall be green; and shall not be careful in the year of drought, neither shall cease from yielding fruit." (Jeremiah 17:8)*

The first thing to notice about this plant is that it is specifically planted by the waters. We do not just grow up wild and by accident. God specifically plants each one of us in a specific place He has chosen for us. For each of us, this is a different and unique place. God has a specific place He wants for us today and tomorrow and each day of our lives. This infinite care for each of us as individuals illustrates clearly that we are each precious in His sight. He knows each of us personally and even gave His life for us, calling us His friends. In doing so, He demonstrated the greatest love of all. *"Greater love hath no man than this, that a man lay down his life for his friends." (John 15:13)*

When heat and drought come to such a plant drawing upon abundant water, they cannot hurt it. In heat, *"her leaf shall be green."* And in drought, she will not *"cease from yielding fruit."*

In the same way, when heat and drought, trials and tribulations, come in our lives, it is the living water of Christ and His Word that sustains us. The Lord echoes this promise to the well-watered plant in Isaiah. *"I the LORD do keep it; I will water it every moment: lest any hurt it, I will keep it night and day."* *(Isaiah 27:3)* The Lord's constant watering makes the plant to withstand even the harshest weather.

Psalm 1 also says, *"And he shall be like a tree planted by the rivers of water, that bringeth forth his fruit in his season; his leaf also shall not wither; and whatsoever he doeth shall prosper."* *(Psalm 1:3)* Again in this verse, the tree is specifically planted by the rivers of water.

One thing to consider is that a tree cannot flee from heat and drought. Once it is planted, it is there to stay. When heat comes, it cannot go into a building with air conditioning. When drought comes, it cannot go to find water elsewhere. It must remain where it is planted.

In the same way, we must remain where God plants us. If God is the One Who plants us, then we need not fear anything that comes our way, for He is able to sustain us where He plants us. The Bible furnishes tragic examples of ones who tried to flee from the place where God had planted them in times of famine. Naomi fled from Bethlehem to Moab and lost her family. Abraham fled from the land of promise to Egypt, which led to lying, almost losing his wife, plagues upon Pharaoh's house, and the loss of a good name before Pharaoh.

These should serve as warnings to us that even when severe heat and drought come, we must not flee from the place of our planting. We must trust the Lord to sustain us where we are, for it is He Who has planted us there. But in order for this to be true, we must make very sure that we are not in the place of our choosing, but in the place where God wants us.

This is a good test of whether or not we are drawing upon God's water for our sustenance. How do we respond to heat and drought? The Bible teaches that these trials and tribulations will come, but they cannot kill or even harm us if we are watered by the Lord. Our leaf will not wither. We will be able to stand up to the stress, not in our own strength, but because Christ has planted us in the right place, at the right time, and sustains us with His living water.

Deuteronomy gives a clearer picture of what is meant by the rivers of water, by which the tree of Psalm 1 is planted. The Lord says in Deuteronomy, *"For the land, whither thou goest in to possess it, is not as the land of Egypt, from whence ye came out, where thou sowedst thy seed, and wateredst it with thy foot, as a garden of herbs." (Deuteronomy 11:10)*

What is this talking about, watering with the foot? It seems strange at the first reading. Does it picture carrying a bucket of water on one's foot? No, instead it pictures making a trench with one's foot in soft soil, directing water from the main irrigation channel into a small trench that carries water directly to the plant. The rivers of water spoken of in Psalm 1 are not the rushing torrents of mighty rivers. Instead, they are these little rivers, specifically brought to specific plants. Not even a shovel is needed to move this water, only the effort of the foot. The leaf of such a plant sustained by the specifically channeled water of God's Word shall not wither. Instead the plant will prosper.

Psalm 1:3 further emphasizes the very individual nature of each plant. It says that the plant brings forth *"**his** fruit."* Christ has a specific purpose for each one of us and specific fruit that He wants us to bring forth. An apple tree cannot produce oranges, but should devote its energy to apples. At the same time, the orange tree should not desire to produce apples, but should devote its energy to oranges. We dare not try to imitate others or bring forth the same fruit that they do. This will lead to failure. Rather we must submit to Christ's unique purpose for our lives and bring forth our own fruit that He has for us.

This verse also says that he brings forth his fruit *"in **his** season."* Just as Christ has specific fruit for each person, He has a specific time for that fruit as well. It will not be too early or too late. It will be exactly right, but different for each person. Christ knows exactly when is the best time for our specific fruit, when it will be the most useful and the most pleasant, and He will cause us to bring forth fruit at that time.

We see in this passage God's infinite concern for us as individuals. He specifically plants us in a specific place, and He gives us specific fruit at a specific time.

Also in this verse we see God's protection. He promises that our leaf shall not wither. God protects us both night and day. It is usually easy to trust the Lord in the day, but His protection is constant. He is just as near to us at night as He is in the day. Though we may not think as well at night or we may be afraid at night, the Lord never changes. He protects us even at night when we are the weakest.

Finally in this verse we see the prosperity that the Lord gives us. He promises, *"whatsoever he doeth shall prosper." (Psalm 1:3)* This is not prosperity as the world means it, but it is spiritual prosperity. But neither is it limited to heaven. The Christian planted by the rivers of water is made to prosper both in this world and the next.

Many Christians may not think that they are prospering now. But that is because they have their eyes on the world's standard of prosperity. Especially in the United States, prosperity has been greatly inflated. Just because we don't have all the wealth that the rich people have, this does not mean that we're not prospering. The poorest people in the United States are in the top three percent of prosperity in the world.

Yet even in third world countries where material wealth is small, Christ is able to give prosperity to those who truly draw upon His living water. Even simple blessings like food to eat and air to breathe are signs of His prospering hand upon us. Otherwise, we would be dead.

The key to true prosperity and true fruit bearing is to draw upon the water that Christ gives, which is the Word of God. As we do, Christ is will be our perfect sustainer, intercessor, and vinedresser.

After water, another necessity for fruit production is that **weeds** be removed. In Mark 4, Christ gives the parable of the seed sown in four different types of soil. We will look at the third kind of soil, the thorny ground. *"And some fell among thorns,*

running into the street, that children who are not given a hedge of protection could lose their lives.

The same is true of young trees. Often we may see a white band around the base of a young tree, or a wrapping of tar paper and cardboard. These are put around the tree to protect it from certain dangers. The bark of young trees is very thin and thus easily damaged by "weed eaters."

It is also in danger of sunscald. The sun penetrates the thin bark and warms the sap even in cold weather. This causes the sap to flow upward. But as soon as the sun disappears, the sap, which is now above the ground and exposed to the cold, freezes. Then the tender bark bursts.

A wire mesh is also sometimes seen around the base of a young tree. This is to protect the tree from mice that chew the thin bark at the tree's base when food is scarce in the winter. When the ground is covered with snow, this chewing can go unnoticed until the snow melts, allowing time for considerable damage to be done. When the tree is young and tender, it cannot withstand this kind of stress. Its bark is not thick enough. For this reason it needs a wire mesh for protection.

But not only the young Christians need protection. Older Christians need to protect themselves as well. As mature Christians, they are beyond the stage where they need others to protect them with wire mesh and paper wrapping. The responsibility for protection is theirs.

This protection is described in Ephesians 6 as the armor of God. *"Put on the whole armour of God, that ye may be able to stand against the wiles of the devil. For we wrestle not against flesh and blood, but against principalities, against powers, against the rulers of the darkness of this world, against spiritual wickedness in high places. Wherefore take unto you the whole armour of God, that ye may be able to withstand in the evil day, and having done all, to stand. Stand therefore, having your loins girt about with truth, and having on the breastplate of righteousness; And your feet shod with the preparation of the gospel of peace; Above all, taking the shield of faith, wherewith ye shall be able to quench all the fiery darts of the wicked. And take the helmet of salvation, and the sword of the Spirit, which is the word of God: Praying always with all prayer and supplication in the Spirit, and watching thereunto with all perseverance and supplication for all saints." (Ephesians 6:11-18)*

We must put on this armor to protect ourselves against the wiles of the devil. This is the eventual goal of protecting the young as well. We should not only protect them, but teach them to protect themselves. Then when they reach maturity, they will be able to stand against the wiles of the devil.

These things are vital to our production of fruit. Thus they are vital to our lives, for fruit production is the purpose for which Christ has planted us and continues to dress and keep us. Let us abide in Him increasingly, that we may increase in fruit production, and may reach the point of maturity where we bring forth much fruit, lasting fruit for the glory of God.

20
The Garden of Eden

At the very dawn of time, plants occupied a prominent place in God's creation. Genesis 2:8 records, *"And the LORD God planted a garden eastward in Eden; and there he put the man whom he had formed."* The word "Eden" comes from the Hebrew root *adan* (עדנ), which means "pleasant, luxury, delight" and pictures the idea of "paradise."

It is interesting to read that the garden was planted **_eastward_** in Eden because most of us have always equated the garden with Eden, and see the two as interchangeable. In reality, the garden occupied only the eastern section of Eden.

It is from the east that God meets man. Centuries later, it was to the east that Solomon's temple was built. It was therefore to the east that the children of Israel prayed. Again centuries later, it was through the eastern gate that Christ entered Jerusalem on a colt.

And in the future, it will be from the east that Christ will return. Matthew 24:27 says, *"For as the lightning cometh out of the east, and shineth even unto the west; so shall also the coming of the Son of man be."*

Ezekiel's vision in Ezekiel 43 also records the glory of God coming from the east, *"Afterward he brought me to the gate, even the gate that looketh toward the east: And, behold, the glory of the God of Israel came from the way of the east: and his voice was like a noise of many waters: and the earth shined with his glory. . . . And the glory of the LORD came into the house by the way of the gate whose prospect is toward the east." (Ezekiel 43:1-2,4)*

Even the sun, the symbol of God's goodness to man, rises in the east. This eastward position, the place where Christ meets man, is the place that was given to the garden in Eden.

As far as the rest of Eden, we are not told what occupied the northern, southern, and western sections. Some have speculated that there was a lower, earthly part to Eden and an upper, heavenly part. This could very well be. A river flowed out of

Eden and into the garden. Since water always flows downhill, this would make it possible for the water to be flowing from the heavenly part of Eden into the earthly part, the garden. I don't know for sure, for the Bible does not say. But what we do know is that the garden occupied the eastern part of Eden.

We also know that, before his sin, Lucifer dwelt in Eden. Ezekiel 28, written as though it were speaking to Lucifer, says, *"Thou hast been in Eden the garden of God; every precious stone was thy covering, the sardius, topaz, and the diamond, the beryl, the onyx, and the jasper, the sapphire, the emerald, and the carbuncle, and gold: the workmanship of thy tabrets and of thy pipes was prepared in thee in the day that thou wast created. . . . Thine heart was lifted up because of thy beauty, thou hast corrupted thy wisdom by reason of thy brightness: I will cast thee to the ground, I will lay thee before kings, that they may behold thee." (Ezekiel 28:13,17)*

The Hebrew word for "garden" is *gan* (גַּן). It is not our normal picture of a garden with rows of peas, radishes, and beans, but it refers rather to an orchard or a park.

To water this green arbor, there flowed a river out of Eden, which parted into four heads. It seems that only one of these four rivers actually flowed through the garden, for it is said of Hiddakel that it went *"toward the east of Assyria." (Genesis 2:14)* Flowing toward the east, it would have flowed through the section of Eden occupied by the garden, but the remaining three rivers seemed to flow in different directions through the other sections of Eden.

The plants that filled the garden were given by the Lord as food for man and animals. Genesis 1 records God's instructions to man concerning the eating of plants. *"And God said, Behold, I have given you every herb bearing seed, which is upon the face of all the earth, and every tree, in the which is the fruit of a tree yielding seed; to you it shall be for meat. And to every beast of the earth, and to every fowl of the air, and to every thing that creepeth upon the earth, wherein there is life, I have given every green herb for meat: and it was so." (Genesis 1:29-30)*

The first thing we must notice about this command is that man was given both herbs and fruits to eat, but animals were to eat only green herbs. Today, however, we see birds eat cherries and squirrels eat nuts. The original diet of only green herbs seems far-fetched to us. But the future diet of animals is equally far-fetched. Isaiah 65:25 says, *"The wolf and the lamb shall feed together, and the lion shall eat straw like the bullock: and dust shall be the serpent's meat. They shall not hurt nor destroy in all my holy mountain, saith the LORD."* We believe this verse that speaks of the millennial kingdom, and so we should also believe Genesis, that it was like this in the beginning.

We must take the Bible literally wherever we can, even though there are a few instances where it is clearly figurative. For example, when Christ says, *"Behold, I stand at the door, and knock" (Revelation 3:20)*, He does not mean He is standing outside a literal door and literally knocking. In the same way, when God calls the earth his *"footstool" (Isaiah 66:1)*, He is not talking about a literal three-legged stool on which to rest His feet. But in cases where the Bible is not clearly figurative, it must be taken literally. Just as Uzzah was killed for violating God's specific instructions on carrying the ark, so are we likewise in danger if we take God's specific and literal Word lightly.

Genesis 1 records God's first creation of plants. *"And God said, Let the earth bring forth grass, the herb yielding seed, and the fruit tree yielding fruit after his kind, whose seed is in itself, upon the earth: and it was so." (Genesis 1:11)* Later, chapter 2 records another planting of plants. *"And the LORD God planted a garden eastward in Eden; and there he put the man whom he had formed. And out of the ground made the LORD God to grow every tree that is pleasant to the sight, and good for food; the tree of life also in the midst of the garden, and the tree of knowledge of good and evil." (Genesis 2:8-9)* Why did God twice make plants to come from the ground? Were they two different types of plants? Were the first plants imperfect? Did God wipe them out, and start all over with new plants?

The key to understanding these two plantings is to notice the words *"earth"* and *"ground."* In the first planting on day three of creation, the *"earth"* brought forth the plants. This first reference in Genesis 1:11 speaks of plants in general over the earth. But the second is more specific since the garden of Eden did not cover the whole earth, but was a particular plot of ground. The second planting in Genesis 2:9 continues the account of the planting of the garden in verse 8 and refers specifically to the planting of the garden of Eden. Here God made the plants to grow out of the *"ground."* These plants were specific to the garden.

Genesis 2:9 says specifically of these plants that they were *"pleasant to the sight, and good for food."* This seems to imply that not every tree on the earth was pleasant to the sight and good for food, but specifically the trees in the garden were. After describing the trees in the garden as such, this verse then separates out two specific trees, saying, *"the tree of life also in the midst of the garden, and the tree of knowledge of good and evil."* (Genesis 2:9) The separate mention seems to indicate that these trees were not part of those that were pleasant to the sight and good for food.

Later, under Satan's deception, Eve would see that the tree of the knowledge of good and evil was *"good for food, and that it was pleasant to the eyes, and a tree to be desired to make one wise."* (Genesis 3:6) But originally she would not have seen these qualities or have been enticed by this tree. In the same way, in our lives there are some things that we may think undesirable, but when we see them in a new light, we see that they indeed are desirable. If we all lived in blue houses, the first green house we saw may seem ugly to us. But after a while, we may realize that it is beautiful and learn to like it. This can either be good or bad. It can either be Satan's blinding of our eyes or the Lord's opening of our eyes. For example, if we at first hate to read the Bible, but then learn to love to read the Bible, this would be of the Lord. It would be His opening of our eyes and would be a good change in our desires.

One more point to notice in Genesis 1 on the creation of plants is the time period of the day. The needs of plants make a strong case for six literal 24-hour days of creation. The plants were created on day three, but the sun, moon, and stars were not created until day four. If, as some say, there were thousands of years between each "day," the plants would have died due to a lack of sunlight. They would have brought forth no fruit due to a lack of seasons. While we know that God could have performed a miracle and preserved the plants over thousands of years and that in Heaven the tree of life lives without a physical sun because Jesus Christ is the sun, still this is not God's way of working on earth. God uses the sun to give growth to the plants and to change seasons so that they produce fruit. The account reads naturally as 24-hour days and should be taken literally as such.

In Genesis 1:3, God said, *"Let there be light."* This **light** was filtered by the firmament so that it didn't penetrate to the earth. In Genesis 1:14-15, God said, *"Let there be lights in the firmament of the heaven to divide the day from the night; and let them be for signs, and for seasons, and for days, and years: And let them be for lights in the firmament of the heaven to give light upon the earth."* These **lights** were certain lights of the light spectrum, such as red, green, and blue light, that benefit plants. This passage clearly states that these lights were *"to give light upon the earth."*

Plants need these lights in order to live since photosynthesis occurs during the daylight, and oxygen is released into the air at night. They need seasons and years for rest in order to continue growth and production; even evergreens need this cycle. In the garden of Eden, these lights were given in perfect balance. Certain individual lights of the spectrum, such as x-rays, gamma rays, and most ultraviolet rays, did not reach the earth because they were harmful. But the daily, monthly, seasonal, and yearly cycles were established and maintained by the beneficial lights.

The charge to care for the garden of Eden was given to Adam in Genesis 2:15. *"And the LORD God took the man, and put him into the garden of Eden to dress it and to keep it."* What did Adam's duties involve? Was Adam to mow, plant, mulch, and prune? No, for there was no sin, so there was no death, no dead branches to prune, no dead grass to mow.

What then does it mean to *"dress it and to keep it"*? The word "dress" literally means to order or to serve, and is the same word used for military order, when troops are commanded to "dress right," which speaks of discipline according to a prescribed order.

Putting this together with the concept of service, the word "dress" means to serve according to a prescribed order. Thus there was an order of service that Adam was supposed to fulfill in the garden. First he was to name the animals and plants. This was no small task, considering the great variety that God had created. When he finished this, no doubt there were other ordered tasks given to him by the Lord, in which he was to serve in the garden.

The word "keep" means to observe, to guard, to look out for, to protect. With the fall of Lucifer, Adam was to guard the garden against the devil. This is why Adam was more responsible than Eve to refrain from eating of the tree of the knowledge of good and evil.

To apply this to the Christian life, we also are to dress and keep our own lives, the gardens with which we have been entrusted. We are to serve Christ according to the order He has prescribed, and are to guard against the wiles of the devil, since Satan has observed Christians for thousands of years, and he is able to be very subtle in his attacks, even transforming himself into an angel of light.

Thus to dress and keep our hearts, we must be attentive to keep God's ordinances and must serve the Lord with all of our heart, soul, mind, and strength.

With Adam's responsibility to dress and keep the garden came also a requirement. God commanded him in Genesis 2:16-17, *"Of every tree of the garden thou mayest freely eat: But of the tree of the knowledge of good and evil, thou shalt not eat of it: for in the day that thou eatest thereof thou shalt surely die."*

Yet alongside this forbidden tree of the knowledge of good and evil, the Lord planted another tree, the tree of life. It first appears in Genesis 2:9, *"And out of the ground made the LORD God to grow every tree that is pleasant to the sight, and good for food; the tree of life also in the midst of the garden, and the tree of knowledge of good and evil."*

The tree of life is symbolic of Jesus Christ. It is mentioned only here and in the book of Revelation. It is *__the__* tree of life, not *__a__* tree of life. *__A__* tree of life, mentioned in the book of Proverbs, pictures certain attributes that we are to have as Christians. Our Christ-like character is to be a tree of life to others, pointing them to Christ, Who is *__the__* tree of life. In John 14:6, Jesus did not say,

"I am *a* way, *a* truth, and *a* life." He said, *"I am **the** way, **the** truth, and **the** life." (John 14:6)* He is **the**. We have *a*.

The tree of life is also symbolic of the cross, as recorded in Acts 5:30, *"The God of our fathers raised up Jesus, whom ye slew and hanged on a tree."* This tree that was a tree of death for Christ is to us the tree of life, whereby we through faith receive life eternal.

The sweet fruit of Jesus Christ, the tree of life, is offered in Psalm 34:8, *"O taste and see that the LORD is good: blessed is the man that trusteth in him."* This is a specific offer that commands us to taste and see **that** the Lord is good, not **if**. It is a sure truth that He is good. We need not taste it to find out if He is good, but only to learn that it is indeed true that He is good.

In matters of food, people have different tastes. Some like certain foods, while others do not. But with Jesus Christ, all have the same taste. If they will only taste, they will see that He is good. The world, which does not know He is good, has simply never tasted.

When I was a child, my parents did not require me to eat everything on my plate. But their rule was that I must at least taste everything. Okra was one new food that was once set on my plate. By sight, the green color and slimy-looking texture made me recoil. But when I tasted it, I found it to be very good.

Like my loving parents, the Lord also requires only that we taste. If we judge Him by appearance, we may conjure up all sorts of misconceptions, that He is too harsh, that He is unjust. We will never know that He is good. But if we will only taste, we will find He is good.

Unlike some foods that taste good, when we taste of Christ, we find that He is also good for us. He both tastes good and is good for our spiritual well-being.

We will look further at the tree of life in future chapters. For now, we leave it where it first appears, in the garden of Eden. Just as the tree of life was in the midst of the garden, Christ is to be in the midst of our lives. There in the garden the tree stands in pristine creation beauty, picturing the eternal life Christ won for us on the cross, and inviting all to *"taste and see that the Lord is good."*

21
Gopher Wood

In Genesis 6:14 God commanded Noah, *"Make thee an ark of gopher wood; rooms shalt thou make in the ark, and shalt pitch it within and without with pitch."* Immediately the question comes to mind, what is gopher wood? No tree is known by that name today.

The word translated "gopher" in the English Bible is in reality not a translation at all, but a transliteration of the Hebrew word *gopher* (גפר). There are a few other words like this in our English Bible, like the word "baptism," which is a transliteration of the Greek word *baptisma* (βαπτισμα).

In the Septuagint, the Hebrew word *gopher* is translated into the Greek *kuparisson* (κυπαρισσον). This is the Greek word for cypress. If the suffix *-isson* is dropped, the word *kupar* is left, which is very close to the Hebrew word *gopher*. The similarity of the words has led many scholars to believe that gopher and *kupar* refer to the same tree, the cypress tree.

Cypress is indigenous to the Middle East area where Noah and his family likely lived and is durable. This fact is used to support their theory that cypress is gopher wood because many claim that the wood for the ark would have had to have been durable. They claim this because of the timing of God's command to Noah to build the ark, not based on the duration of the Flood.

The duration of the Flood was only about a year. Genesis 7 says the rain began *"In the six hundredth year of Noah's life, in the second month."* *(Genesis 7:11)* And Genesis 8 says that the earth was dried a year later, *"in the six hundredth and first year, . . . in the second month."* *(Genesis 8:13-14)*

But though the ark floated only about a year, some say it took 120 years to build. They say this because Noah preached and warned for 120 years. This was the amount of time God gave the wicked men of that generation to repent. *"And the LORD said, My spirit shall not always strive with man, for that he also*

is flesh: yet his days shall be an hundred and twenty years."
(Genesis 6:3)

But this doesn't necessarily mean that Noah was actually building the ark for all those years. It means only that he was a witness for that long. Evidence shows that the men of that day were highly skilled craftsmen. For such skilled men as Noah and his sons, it wouldn't have taken 120 years simply to build the ark. Also, wood was easier to work than stone, which they normally worked. So they would have easily been able to complete the ark in a short amount of time.

We must remember also that there was not a long period of time between the completion of the ark and the flood. Noah didn't have to go out and collect all the animals Instead God sent them, and God's work certainly didn't take very long.

So at the very most, the wood of the ark would have had to last for a few years of building and then for another few years while the animals came. This would be the very most, for the animals could very well have been gathered at the same time the ark was being built, since God gathered them and not Noah.

The Bible doesn't say exactly when during the 120 years God told Noah to build the ark. After the 120 years were pronounced, in the same chapter, we read, *"But Noah found grace in the eyes of the LORD." (Genesis 6:8)* He probably was a witness to the ungodly generation about him during the early part of the 120 years, long before God told him to built the ark. For that command is not recorded until verse 14 of the chapter, *"Make thee an ark of gopher wood; rooms shalt thou make in the ark, and shalt pitch it within and without with pitch." (Genesis 6:14)*

So though cypress wood is durable, this is not necessarily a requirement for the ark. It did not necessarily have to last a long time. Other reasons that some call gopher wood cypress wood are that it is water resistant and decay resistant, but so are many other trees. We also know that cypress needs a lot of sunlight in order to grow, but since we don't know the weather conditions before the Flood, it is almost impossible to know whether the cypress would have grown abundantly or not.

In spite of the good arguments for gopher wood being cypress wood, there are problems with drawing this conclusion. Cognates between two languages are not conclusive. The Old Testament was written in Hebrew, not Greek. The original word is *gopher*, not *kupar*.

If the Lord had meant cypress wood, there was a Hebrew word for cypress available to Him. This is the word *tirzah* (תרזה), used in Isaiah 44:14, *"He heweth him down cedars, and taketh the **cypress** and the oak, which he strengtheneth for himself among the trees of the forest: he planteth an ash, and the rain doth nourish it." (Isaiah 44:14)* Since the Lord chose to use the word *gopher*, rather than *tirzah*, it seems that He meant something other than cypress wood.

There are other kinds of wood that would serve the purpose of the ark. The British navy built their ships out of white oak wood. White oak, like the cypress, is water resistant. Red oak, on the other hand, is filled with hollow tubes that easily allow water in. In fact, you can blow on one end of a piece of red oak and feel air come out the other end. But white oak is very resistant to water, and is still used today in whiskey barrels.

Another wood that might have served Noah's purpose is the Australian Gray Ironbark. This tree is a kind of eucalyptus that was used for making the prow of arctic vessels in order to break ice and clear a path for the vessels. It is metal-like and therefore hard to fashion, but very strong and durable. It can last hundreds of years if it is properly treated. But even this I don't believe to be gopher wood.

We have not yet examined the term "gopher wood," but simply ruled out cypress wood for the ark. We have done so pretty easily. Cypress is not necessary to the ark, for other woods serve the purpose equally well. Also, if God had meant cypress, it seems He would have used the word *tirzah*, as He did in Isaiah. But instead He used the word *gopher*.

Now we will look at the term "gopher wood" itself. Some scholars have suggested that the word *gopher* may have been mis-scribed. The letter "g" is very similar to the letter "k" in the Hebrew alphabet. If the word were *kopher* instead of *gopher*, it would be the word used in the latter part of Genesis 6:14 for pitch.

But I don't believe this is the case, since God promises to preserve His Word. Christ said, *"For verily I say unto you, Till heaven and earth pass, one jot or one tittle shall in no wise pass from the law, till all be fulfilled." (Matthew 5:18)* If this word were mis-scribed, what else in the Bible might have been mis-scribed?

Copying of the Bible has been done very carefully over the centuries. Copies have been checked and rechecked in dozens of ways. When the Dead Sea scrolls were discovered in 1930, the book of Daniel found among them was compared with newer copies. It matched almost exactly word for word. Copying of the Bible over the years has been very accurate, according to God's promise.

Even if this word were mistranslated, the verse would not make any sense. It would read, "Make thee an ark of **pitch**; rooms shalt thou make in the ark, and shalt **pitch** it within and without with **pitch**." The Septuagint translates "pitch" as either *asphaltos* (ασφαλτος) or *zephet* (ζεφετ), clearly a sticky, tar substance. It would be impossible to make a ship entirely of this pitch, and it would also be illogical to pitch the pitch with pitch.

It is somewhat instructive to look at the various ways that "gopher" has been translated. The Septuagint LXX translates it "squared beams." The Latin Vulgate translates it "planed wood." The word "gopher" is defined in the Concise Oxford Dictionary as wafer. In a wafer there are layers, for example, a cookie or wafer, then icing, then another cookie, then more icing.

Still the words are similar. One letter changes gopher to *kopher*. There seems to be a close relationship between gopher wood and *kopher*, pitch. Perhaps then gopher wood is not a kind of wood, but a process done to wood. Today we have plywood, but there is no ply tree. This would explain the need for pitch both within and without. This is both within and without of the ark, and also within and without of the wood. The pitch would actually be in the wood as a lamination, like plywood or LVL beam, and would form it into a wafer, making gopher wood.

Dr. John Hinton from Harvard supports this theory by saying that this pitch was not a petroleum product. He says that the word "pitch" means any waterproof covering, probably in this case the resin from a tree or plant.

Kopher, the word for pitch, is also the Bible word for covering, or atonement. Just as the pitch was to be applied both within and without and provided a covering for Noah to go through the Flood, we must have Christ's atonement applied to our lives. We must be changed both inside and outside. The atonement must be internal, actually within the very fiber of our hearts, just as the pitch was within the fiber of the wood. If there is no change on the outside, it is evidence that there was no change on the inside.

This brings up the question of whether or not the men of ancient days could do lamination. One interesting fact to note is that long before plywood as we know it, American Indians were using "pitch" from pine trees and fir trees to glue arrows and to do other wood laminating. This knowledge of lamination has at least been around prior to modern times.

There is also evidence that before the Flood, men were not only as advanced as today, but that they even had knowledge of things we don't today. Archeologists have dug up axes, hammers, and spears with the wooden handles glued to the metal or flint heads by a very strong bituminous material. This strong glue is still holding today, these thousands of years later. It must have a much stronger bond than that produced by any super glue we have today.

In 1929 Sir Leonard Wooley uncovered the lyre of Ur, a lyre (harp) made with laminated wood in Ur of the Chaldees. He also found plywood boxes made in 3500 B.C. buried in Egyptian tombs dated 2000 B.C.

Another evidence of the advanced culture of the pre-Flood world is the amazing accuracy of the pyramids. Some of the stones are so precisely placed that one cannot even slip a piece of paper between them. With our advanced engineering technology, we cannot approach this mathematical accuracy. By no means were the ancient men dummies, far from it.

Based on the Biblical measurements, the ark would have been 450 feet long, 75 feet wide, and 45 feet high. The *HMS Victory*, the last and largest wooden ship built, was only 226 feet long, 51 feet wide, and 21 feet high. This means that the ark would have had six times this ship's volume and would have used five times its amount of wood.

Made of 95 percent white oak, the *Victory* used up 6,000 oak trees between 12 and 30 inches in diameter. Thus the ark would have used 30,000 trees of the same size. It is perfectly reasonable to believe there were that many trees before the Flood, since the growing conditions were ideal.

Over the years, many people have made models of the ark in order to understand it better. While this is helpful when the Biblical pattern is followed, it becomes dangerous when men impose their own ideas upon the pattern. For example, some have proposed that the ark had a fin protruding into the air to catch the wind and allow the boat to be steered. Others have proposed that steering was accomplished with a rudder. Still others have proposed that the ark had an anchor to keep it in a small area.

But none of these things are found in the Bible. They are only manmade ideas reflecting what man thinks the ark should have looked like. They serve only to remove God from the picture and to deny that He was the One Who steered the ark, Who decided where it would eventually rest. God could put the ark wherever He wanted it, anchor or no anchor, rudder or no rudder, fin or no fin.

The ark is a type of the nation of Israel in the last days. Just as God took Noah through the Flood, He will take Israel through the Tribulation. But for the Christian, Enoch is the type. Just as God translated Enoch before the Flood, so He will translate Christians before the Tribulation.

A question that arises when talking about the Flood is how plants were dispersed afterward. There are three main ways that seeds are carried, animals, wind, and water. Animals eat the seeds and later drop them with their waste in a new location, sometimes very far away. Some seeds are carried by the wind, particularly seeds like the milkweed that are specially equipped for air travel. Water also carries seeds. Coconuts have been known to survive a sea journey of several months and to begin new coconut groves on distant shores.

It is entirely possible and very common for seeds to remain in the ground without germinating, waiting for right conditions. Sometimes years later, temperature, moisture, or light conditions change, and these seeds germinate and sprout into plants. Such could have happened to seeds that went through the Flood. Waiting dormant in the ground throughout the year of the Flood, they would have sprouted the next year when conditions were right.

Seeds have been known to survive dormant for several years, even thousands of years, and still be viable when given right conditions. In Herod's tomb, for example, ancient seeds at least two thousand years old were discovered that germinated when they were planted. The same thing happened to two-thousand-year-old date seeds found hidden in the caves with the Dead Sea scrolls.

There is much we can learn from the ancient, pre-Flood world. While most commentaries say gopher wood is cedar, cypress, or juniper, this is only because they are resinous. Instead I believe the evidence points to gopher wood being a laminated wood, stronger than any we have today.

The men of Noah's day had a far more advanced civilization than we. Yet they decayed into great wickedness, such that God saw *"that the wickedness of man was great in the earth, and that every imagination of the thoughts of his heart was only evil continually." (Genesis 6:5)* If the men of Noah's day could fall into such wickedness, certainly we can as well. Like Noah, we need an atonement that will change us both within and without, and will bring us safely through this world and into the next.

22
Grafting/Firstfruit

Romans 11:16-25 presents a vivid picture of the Gentiles as a wild olive branch grafted into Christ. But this passage sounds a bit confusing at first, for some branches are broken off, while others are grafted in. So we will take it verse by verse in order to discern carefully its meaning.

Verse 16 opens the discussion with a reference to the firstfruit. *"For if the firstfruit be holy, the lump is also holy: and if the root be holy, so are the branches." (Romans 11:16)* Throughout the Bible, the firstfruit is usually the best fruit. The reason to graft a plant into another plant is to produce a different result. The Gentiles in this passage are grafted into the root, the best part, to produce a better result for Christ.

Numbers 15 commands that the firstfruit of bread dough, the first and best part, be offered as a sacrifice to Christ. *"Then it shall be, that, when ye eat of the bread of the land, ye shall offer up an heave offering unto the LORD. Ye shall offer up a cake of the first of your dough for an heave offering: as ye do the heave offering of the threshingfloor, so shall ye heave it. Of the first of your dough ye shall give unto the LORD an heave offering in your generations." (Numbers 15:19-21)*

Malachi applies the same principle to tithing. *"Bring ye all the tithes into the storehouse, that there may be meat in mine house, and prove me now herewith, saith the LORD of hosts, if I will not open you the windows of heaven, and pour you out a blessing, that there shall not be room enough to receive it." (Malachi 3:10)* It is God Who deserves the first and best of our wealth. If we render the firstfruit to Him, then He will abundantly bless the remainder.

Leviticus also lays down the principle that the firstfruit is the best. *"Speak unto the children of Israel, and say unto them, When ye be come into the land which I give unto you, and shall reap the harvest thereof, then ye shall bring a sheaf of the firstfruits of your harvest unto the priest:" (Leviticus 23:10)* This was to be the first generation of crops harvested in the land of Israel, and therefore was to have the most vigor and the most nutrients from the soil.

In our text of Romans 11:16, the firstfruits referred to are the Jewish people. *"For if the firstfruit be holy, the lump is also holy: and if the root be holy, so are the branches." (Romans 11:16)* The Jewish people are holy branches on the holy root of Christ.

Yet Jeremiah 11 records their sin of idolatry and the harsh judgment of being "broken off" from the Lord as a branch from a tree. *"Therefore pray not thou for this people, neither lift up a cry or prayer for them: for I will not hear them in the time that they cry unto me for their trouble. What hath my beloved to do in mine house, seeing she hath wrought lewdness with many, and the holy flesh is passed from thee? when thou doest evil, then thou rejoicest. The LORD called thy name, A green olive tree, fair, and of goodly fruit: with the noise of a great tumult he hath kindled fire upon it, and the branches of it are broken. For the LORD of hosts, that planted thee, hath pronounced evil against thee, for the evil of the house of Israel and of the house of Judah, which they have done against themselves to provoke me to anger in offering incense unto Baal." (Jeremiah 11:14-17)*

Israel had committed such sin in offering incense to Baal that Jeremiah was commanded not even to pray for them. God would not hear prayers for a nation that had turned its back upon Him. The finality of their judgment is seen in that their branches would be broken. Once Israel's branches were broken off, it would be humanly impossible to graft them back in again, because it takes new, green branches for grafting, not old, broken ones.

Yet this is not impossible with God, for God can graft in even old branches. And even on an old tree there may be young and tender branches. To the individual Jewish person, God's mercy is yet extended. Verse 23 of Romans 11 declares God's ability to do the impossible for these broken branches, *"And they also, if they abide not still in unbelief, shall be graffed in: for God is able to graff them in again." (Romans 11:23)* Just as a piece of dough broken off from a lump can be worked back into the dough while it is still soft, a tender branch can be grafted back into the tree.

God's mercy continues to be seen as we move into verse 17. *"And if some of the branches be broken off, and thou, being a wild olive tree, wert graffed in among them, and with them partakest of the root and fatness of the olive tree." (Romans 11:17)* Only some of the branches were broken off. Thus some were left, for it is *"with them"* that the Gentiles partake of the olive tree. The Jewish people are by no means hopelessly "cut off," but can be saved by partaking of the same *"root and fatness of the olive tree,"* Jesus Christ. He is the root and trunk of the natural olive. In Him, both the natural branches, the Jewish people, and the wild olive branches, the Gentiles, may partake of His root and fatness.

The word "graft" means to prick, to cut, or to slice. The tree has to be cut or sliced in some way, the branch placed into the cut, and then they must be bound together with twine. This is a beautiful picture of Christ, who was pierced on the "tree," or "cross," and then bound with linen clothes so that we, both Jews and Gentiles, might be grafted in.

Galatians 3 shows the unity of Jew and Gentile in Christ. *"That the blessing of Abraham might come on the Gentiles through Jesus Christ; that we might receive the promise of the Spirit through faith." (Galatians 3:14)* It is the blessing of Abraham, the blessing promised to the Jewish people, yet it comes on the Gentiles. All of this is accomplished through Jesus Christ in Whom *"there is neither Jew nor Greek, there is neither bond nor free, there is neither male nor female: for ye are all one in Christ Jesus. And if ye be Christ's, then are ye Abraham's seed, and heirs according to the promise." (Galatians 3:28-29)* This unity was Christ's plan all along.

While many parallels can be drawn between man's literal grafting of trees and Christ's grafting of men, there is an important difference between man's grafting and God's grafting. In Bible times, men would graft a good wild olive branch into a dying olive tree in order to make the tree better. Greece, the number one producer of olives and therefore experienced at olive grafting, still uses this practice today.

Yet while man grafts to make a better ***tree***, Christ grafts to make a better ***branch***. We graft in order to get better fruit, a taller tree, a shorter tree, better flowers, better color, or whatever qualities we desire in a certain tree. But Christ already is the perfect tree and the perfect root. His reasons for grafting and His way of grafting are totally different from ours and are infinitely better than ours. He grafts to make us, the branches, better.

Grafting is skilled work. It requires someone who knows what he is doing. Yet even with highly skilled nurserymen, man's grafts sometimes don't take. If this happens and the grafted branch dies, the tree returns to its old state of weakness and is unproductive, rather than being infused with new life. While this may happen with man's grafts, God's grafts always take. He is the Master of all trades, and when He grafts a life into Himself, that life is secure and assured of acceptance and sustenance from the root, Jesus Christ.

Even the old grafts which are impossible with man are possible with God. It is hard for the old broken branches, the Jewish people, to be saved. But they can be when God grafts them in, *"for God is able to graff them in again." (Romans 11:23)*

There is only one principle that must be followed in grafting. Any person, Jew or Gentile, man or woman, can be grafted into Christ if this one principle is followed. The principle is this: the branch grafted in must be similar in nature to the tree into which it is grafted. Two dissimilar trees cannot be grafted into one another.

For example, a branch from a red delicious apple tree can be grafted into a yellow delicious apple tree. But a branch from an orange tree cannot be grafted into a yellow delicious apple tree. In the same way, the branch that is to be grafted into Christ must be born of Christ. He must have the very life and nature of Christ. In other words, he must be saved. Only then can he be grafted into Christ.

Once grafted into Christ, he draws upon Christ's root and fatness. He produces more fruit. He is more colorful, more appealing to the outside world, which sees the change and knows that the grafted branch has something it doesn't have.

The wild olive tree spoken of in Romans 11 is an invasive species. When introduced into an area, it spreads rapidly through its roots. It doesn't take long before it takes over the area. This is the way it is in the world today. The wild olive trees, the unsaved, have filled the world like an invasive species. They have invaded every corner of the world, choking out the indigenous plants, and taking over society with sin and decay. This great abundance of lost people in the world makes the need for grafting all the more pressing. All of these wild olive branches need to be grafted into Christ.

Verses 18 and 19 of Romans 11 go on to warn against pride for those who have been grafted in. *"Boast not against the branches. But if thou boast, thou bearest not the root, but the root thee. Thou wilt say then, The branches were broken off, that I might be graffed in." (Romans 11:18-19)* We dare not boast, saying that the Jewish people were broken off just so **we** could be grafted in. Rather, we would do well to remember that we are the inferior species. We are wild olive branches, not the pure natural branches, which are far more important. The grace of Christ, in saving us and grafting us in, was given not to make us proud, but to make us thankful. Rather than boasting, we should humbly thank the Lord for the privilege of partaking with His people of His root and fatness.

The very fact that we have been grafted, not born, into Christ should make us humble. We are not naturally born Christians. Just because we are born in America or just because we have Christian parents, this does not automatically make us Christians. We are instead all born sinners. We are but wild olive branches by nature.

As branches, we are powerless to help ourselves. No wild olive branch can graft itself into a natural olive tree. We dare not think that we can obtain a place in Christ by our own works either, for only Christ can graft us into Himself.

Yet the reverse is also true and gives us great assurance. While we cannot graft ourselves into Christ, neither can we "un-graft" ourselves out of Christ. We thus cannot lose our salvation. Only Christ can graft us in, and He keeps us eternally secure.

In order that we might have this eternal security that comes through grafting, Christ had to be wounded. Before a branch can be grafted into any tree, the tree must first be wounded. On the cross of Calvary, sometimes referred to as the "tree," Christ suffered this wounding for us. Without it, we could not be grafted in. But because *"he was wounded for our transgressions," (Isaiah 53:5)* we can now be grafted into Him, partake of His life, and find eternal security.

We must be careful, in drawing this analogy, that we do not carry it too far to the individual. When a graft is made, the tree is wounded only once, usually at its base, and a branch is grafted in. Multiple wounds would harm the tree. In the same way, Christ was wounded only once for all mankind. Each time someone is saved, Christ does not have to be crucified again. He was crucified once, and through that one wound all men are offered the privilege of being grafted in to His root and fatness. The wound must be personally applied to each individual. But the wound is made only once, and the soul is saved only once.

This wounding of Christ is pictured in Colossians as the circumcision of Christ. *"In whom also ye are circumcised with the circumcision made without hands, in putting off the body of the sins of the flesh by the circumcision of Christ." (Colossians 2:11)*

In the presence of this Christ, we dare not boast. The greatness of Jesus Christ, the natural olive tree, should produce in us true humility. Even the tree branches at His triumphal entry bowed down at His presence. *"And many spread their garments in the way: and others cut down branches off the trees, and strawed them in the way." (Mark 11:8)* Like these branches, we should fall down before Him and worship Him in true humility.

Verse 20 of Romans 11 warns further, *"Well; because of unbelief they were broken off, and thou standest by faith. Be not highminded, but fear." (Romans 11:20)* It was not because of any act that the Jewish people committed that they were broken off. Though they were often guilty of acts of complaining, this was not what caused their ultimate breaking off. Rather it was their unbelief. When Christ came unto His Own, His Own received Him not. When the Jewish authorities crucified Christ, they were revealing their unbelief, their failure to believe in Him as Savior.

General belief differs from specific faith. Faith in Christ is imparted by God. Romans 10:17 teaches, *"So then faith cometh by hearing, and hearing by the word of God."* However a general belief in God is not saving faith. James says it is possible even for devils to believe generally in God. *"Thou believest that there is one God; thou doest well: the devils also believe, and tremble."* *(James 2:19)* The devils may believe, but they don't have faith. Hebrews teaches that it is only this specific faith in Christ that pleases God. *"Now faith is the substance of things hoped for, the evidence of things not seen. . . . But without faith it is impossible to please him: for he that cometh to God must believe that he is, and that he is a rewarder of them that diligently seek him."* *(Hebrews 11:1,6)*

The general spirit of unbelief of the Jewish people and their specific rejection of Christ as Savior caused them to be broken off from the olive tree. But before we become high-minded, let us consider our own vulnerability. Verse 21 of Romans 11 warns, *"For if God spared not the natural branches, take heed lest he also spare not thee."* *(Romans 11:21)* If they were broken off for unbelief, what of us? We are the wild olive, not the natural branches. Would it take less than unbelief for us to be broken off? Being already a foreign part of the tree, if we fail to produce fruit, how much easier would it be for Christ to cut us off?

History teaches that it is easy for God to cut off whole nations that turn their backs on Him. Throughout history, the nations that have honored the Law of God, that have operated by Biblical principles and laws, have prospered and flourished under God's blessing. But those that look to their own reason and seek their own pleasure have been completely cut off from God's blessing.

We focus much attention and place much confidence in our military. But the military, as important as it is, provides no national security apart from the blessing of God. As evidenced by the nation of Israel, a nation that honors God can be victorious in battle even against overwhelming odds. Thus the biggest threat to the United States is not al-Qaida or any other group, but the turning of our own backs upon Christ. For national prosperity, we dare not trust our military, but rather Christ.

If we do not want to be cut off as a nation, we must draw nigh unto God. He promises, *"Draw nigh to God, and he will draw nigh to you." (James 4:8)* How we regard Christ is how He will regard us. If we turn our backs on Him, He must turn His back on us. But if we draw nigh to Him, He promises He will draw nigh to us.

For the Gentiles, a different mode of removing them from the olive is used. Romans 11:22 says, *"Behold therefore the goodness and severity of God: on them which fell, severity; but toward thee, goodness, if thou continue in his goodness: otherwise thou also shalt be cut off." (Romans 11:22)* While the Jewish nation was **_broken_** off, the Gentiles stand in danger of being **_cut_** off. The word translated "cut" is ekkopto (ekkoptw).

It means to sever, to hack with a knife, or to chop. To be a Gentile and be allowed access to Christ and then to reject Him, is to bring upon oneself a harsh judgment, that of being severed from Christ. The closer we are to Christ, the greater is our responsibility. If we are as close to Christ as a grafted branch and we reject Him, God must sever us from the tree.

To be removed from Christ is a severe judgment as this verse states. Only here in all the Bible is the word "severity" used, both in English and in Greek. At its root is the very word "sever." It carries with it a note of finality, picturing a permanent severing from Christ.

John 15 presents the picture of Christ as the vine. It speaks of pruning the unfruitful branches, carefully cutting them off the vine. *"I am the true vine, and my Father is the husbandman. Every branch in me that beareth not fruit he taketh away: and every branch that beareth fruit, he purgeth it, that it may bring forth more fruit." (John 15:1-2)* Christ doesn't want an unprofitable branch. He wants one that is profitable to Him, one that uplifts and glorifies Him. Therefore He will cut off any that does not bear fruit. Such will be the fate of all who fail to continue in Christ.

However even in this judgment, there is mercy. A branch that is cut smoothly from a tree has a surface for good contact. It can be grafted back in much more easily than a broken branch.

Yet even for the broken branches, the Jewish nation, verse 23 of Romans 11 speaks of God's ability to graft them back in. *"And they also, if they abide not still in unbelief, shall be graffed in: for God is able to graff them in again." (Romans 11:23)* It is nearly impossible for man to graft a broken branch onto a tree. But Christ is able to, for *"with God all things are possible." (Matthew 19:26)*

Even in the human realm, it is sometimes possible to graft a broken branch back into the place from which it broke. The broken piece fits perfectly with the wound in the tree, and thus there is good contact. There can be successful grafting if the broken branch is carefully placed in the wound and wrapped with grafting tape within just a day or two after it breaks.

But this is only possible for the first few days. A rose bush might wait five days and still receive the graft. But any longer and the plant will begin to compartmentalize, cutting off that broken section from receiving nourishment. It would certainly be impossible to graft the branch back in after two thousand years. Yet that is what God promises to do for the Jewish people, who have been broken off for two thousand years.

It must be remembered, too, that *"one day is with the Lord as a thousand years, and a thousand years as one day." (II Peter 3:8)* To the Lord, the Jewish people have been broken off for only a few days. He is fully able to graft them back in, just as a nurseryman would graft in a branch that been broken for a few days.

The passage in Romans goes on to assert Christ's full ability to graft His people back in. *"For if thou wert cut out of the olive tree which is wild by nature, and wert graffed contrary to nature into a good olive tree: how much more shall these, which be the natural branches, be graffed into their own olive tree?" (Romans 11:24)* The process of grafting is found only here in all the Bible. Again in this verse we see the difference between man's grafting and God's grafting. The olive tree is good and perfect. God's grafting is done for the benefit of the branches, not the tree.

The wild olive tree mentioned in this verse is unfruitful apart from Christ. Literal wild olive trees are also unfruitful. They bear olives, but the olives are barren and cannot reproduce new olive trees. Wild olive trees propagate themselves through their roots or through cuttings placed in the soil.

In the same way, we, the wild olive branches, are unfruitful. We must be grafted into Christ, the natural olive, so that we can become productive and truly fruitful.

This benefit is given only to those who have met the condition at the opening of the verse, *"if thou were cut out."* Not all Gentile branches have met this condition. Some branches are not cut out of the wild olive and grafted in to Christ. The Bible is clear that some Gentiles are grafted in, or saved, but some are not grafted in, or are unsaved. *"Enter ye in at the strait gate: for wide is the gate, and broad is the way, that leadeth to destruction, and many there be which go in thereat: Because strait is the gate, and narrow is the way, which leadeth unto life, and few there be that find it."* (Matthew 7:13-14)

By being grafted into Christ, we become a part of His very being. By partaking with the natural branches, we also become a part of Abraham's seed, as Galatians says. *"And if ye be Christ's, then are ye Abraham's seed, and heirs according to the promise."* (Galatians 3:29) In a sense we are "Jewish," spiritually speaking. For *"with them* [the Jewish people]*"* we partake of the root and fatness of the olive tree.

The final verse we look at in this passage in Romans 11 is verse 25. *"For I would not, brethren, that ye should be ignorant of this mystery, lest ye should be wise in your own conceits; that blindness in part is happened to Israel, until the fulness of the Gentiles be come in."* (Romans 11:25) In this verse we see that there will be an end to the time of the Gentiles. For now, the Jewish people are partially blind. It is not total blindness, for only some of the branches were broken off. But when the Gentiles become unfruitful, when their fullness is come in, the Jewish people as a whole will be grafted back in.

God will do this for His people in mercy, not because they deserve it, but for the sake of His faithful servants, patriarchs like Abraham, Moses, and David. For their sake, Israel will be saved and preserved.

Before we close this section on grafting, we look briefly at the characteristics of the olive tree itself. God used it for a reason to illustrate the principle of grafting.

First it is an evergreen. It pictures the everlasting nature of the Jewish nation. Though some branches are broken off, the Jewish people are never completely cast away. They are only partially blind and will one day be wholly grafted back in.

The olive is also a symbol of peace and victory. This comes from the Biblical account of Noah. When the Flood waters were receding and Noah released a dove from the ark, it returned with an olive branch, symbolizing that God had gained the victory over sin and that the world was once again at peace with God.

When men have an argument and one attempts to reconcile with the other, it is said that he "extends the olive branch." He offers peace once again. In the same way, Christ, the olive tree, offers us peace with God, a cessation from striving against Him, by being grafted into Himself.

The olive fruit also exhibits a characteristic that can be translated to the Christian life. It must be put under pressure in order to release its oil. In the same way, we as Christians must be put under pressure before our lives release the sweet savor of oil. When things are going well, we tend not to think about Christ. We forget that it is His blessing we enjoy, and we think that we have the power to sustain ourselves. For this reason, we must be put through trials and tribulations. It is then that we learn to trust Christ as our sufficiency.

The olive tree is mentioned about thirty times in the Bible. It is an invasive species because it propagates itself through its roots. One kind of olive, the Russian olive, is taking over certain parts of the United States. Even when one tree is cut down, two more may spring up from the roots.

The olive is very hardy and resistant to both insects and disease. It is a hard wood and difficult to cut. But when it is cut, it is good wood for burning, producing two and a half times the amount of heat as oak wood. It also burns cleaner, putting less pollutants into the air.

The garden of Gethsemane, where Christ prayed before His crucifixion, abounded with olive trees. The very name "Gethsemane" means "oil press." It was here that olives were put under pressure and their oil released, just as Christ, under intense spiritual pressure, poured out His soul for our sakes.

Finally, it is fitting that the olive tree originated in Israel. Like the Jewish nation it pictures, its native home is Israel, the place of God's choosing. We, the grafted branches, have the blessed privilege of partaking with the Jewish people of their Messianic blessing. We are become part of *"Abraham's seed, and heirs according to the promise." (Galatians 3:29)*

23
The Hazel Tree

The hazel tree is mentioned only once in the Bible. *"Jacob took him rods of green poplar, and of the hazel and chesnut tree; and pilled white strakes in them, and made the white appear which was in the rods." (Genesis 30:37)*

First we must say that this is a true hazel. Though many say it is really an almond, the almond is already mentioned elsewhere in the Bible. For example, Aaron's rod was made from an almond tree. The hazel must be something distinct from the almond.

One use of the hazel was as a divining rod. One of its branches would be taken from the tree, peeled, sharpened to a point, and then used to divine water. It was also used superstitiously to divine criminals and point out the guilty one. Whether or not Jacob used the hazel superstitiously, we don't know. But we do know that God blessed him, and this is strong evidence that he was not dabbling in superstition.

The Hebrew word for hazel is *luwz* (לוז), also spelled *luz*. This was the Hebrew name for the city of Bethel before Jacob named it Bethel, "the house of God." Jacob *"called the name of that place Bethel: but the name of that city was called Luz at the first." (Genesis 28:19)* Probably this city was originally named Luz because there were many hazel trees in its vicinity.

We must remember that the climate of Israel was totally different in the days of the Bible than it is now. Israel is barren today and couldn't sustain such plant life, but it used to. Jericho was once a well-watered city full of palm trees, though today it is a barren wilderness, nowhere near being the lush plain that it once was. But such was the environment in which the hazel tree once flourished in Bible times.

As Jacob learned at Luz, or Bethel, the hazel tree teaches that it is vain to trust in our own superstition, reasoning, and inclination. Rather, we must look to the Lord alone for "pointing out" the way we must go. As we seek Him, He will meet us as He did for Jacob, and turn our Luz into Bethel, "the house of God."

24
The Heath

The heath is mentioned only twice in Scripture, and only in the book of Jeremiah. Its first reference is in Jeremiah 17, where it is used to prophesy against the unfaithful men of Judah. *"Thus saith the LORD; Cursed be the man that trusteth in man, and maketh flesh his arm, and whose heart departeth from the LORD. For he shall be like the heath in the desert, and shall not see when good cometh; but shall inhabit the parched places in the wilderness, in a salt land and not inhabited." (Jeremiah 17:5-6)*

An important key to note in this passage is the placement of the heath. It is *"in the desert,"* inhabiting *"the parched places in the wilderness,"* and *"in a salt land."* We will see the significance of these phrases later as we begin to examine the characteristics of the heath.

The next few verses go on to contrast this dry heath with a well-watered tree. Whereas the heath pictures the man whose heart departs from the Lord, the next tree pictures the man who trusts in the Lord. *"Blessed is the man that trusteth in the LORD, and whose hope the LORD is. For he shall be as a tree planted by the waters, and that spreadeth out her roots by the river, and shall not see when heat cometh, but her leaf shall be green; and shall not be careful in the year of drought, neither shall cease from yielding fruit." (Jeremiah 17:7-8)* While the heath, the unfaithful, was planted in a dry land, the trusting one was *"planted by the waters."*

A similar reference appears later in Jeremiah, in his prophecy against Moab. *"Flee, save your lives, and be like the heath in the wilderness. For because thou hast trusted in thy works and in thy treasures, thou shalt also be taken: and Chemosh shall go forth into captivity with his priests and his princes together." (Jeremiah 48:6-7)*

To understand these prophecies fully, we must first know something of the characteristics of the heath. It is a small evergreen shrub. Some commentaries say that it is the same as a juniper. But I Kings records that Elijah sat under a juniper tree.

This would be very difficult if the juniper were the same as the heath, because the heath gets no taller than three feet. The most it could shade would be Elijah's feet. So the heath must be something different from the juniper.

Heath also brings to mind the Scotch heather because of the similarity of their names. And indeed these two plants are very similar. Both grow in mild climates and in similar soil conditions. But they are not the same. Heath has wider leaves than heather's needle-like leaves. Also heath comes in more colors than heather. But for practical purposes, you can picture heather to get a picture of heath.

The Hebrew word for heath is *arower* (עַרְעָר). It grows in mild climates, not in extreme heat or cold. It certainly doesn't grow in the wilderness or desert. Yet this is exactly where it is said to be, *"in the desert"* in Jeremiah 17, and *"in the wilderness"* in Jeremiah 48. Thus the point of these passages is that the unfaithful men of Judah and the sinful nation of Moab are out of place.

The heath is also said in Jeremiah 17 to be in *"the parched places in the wilderness."* Yet the heath likes lots of water. It is further said to be *"in a salt land."* This is alkaline land. But heath is an acid-loving plant. It grows best in acidic soil. On the Ph scale, acidity is on the exactly opposite end of alkalinity. The heath of Jeremiah 17, placed in *"a salt land,"* is placed in a land exactly opposite to its natural conditions.

On all counts, this heath is out of place. It is in a hot place, a dry place, and a salt land. It is in a place where it does not normally grow.

Throughout this book, we have come across a few other plants that were out of place. In chapter 9, we saw that Balaam likened the children of Israel to *"cedar trees beside the waters."* *(Numbers 24:6)* Beside water is out of place for a cedar tree. But at the time of Balaam's prophecy, Israel was wandering in the wilderness and had not yet arrived in the promised land. They were out of place, and thus the out-of-place cedar trees were a fitting analogy. We also saw in chapters 7 and 18 that the Lord planted the fir, pine, and box trees in the desert, not their normal habitat.

Both of these occurrences show that the Lord is all-powerful. He can plant and sustain trees, and people, in any conditions He chooses, even conditions that are exactly opposite to their normal needs. When the Lord is the One Who plants the tree, even the most harsh location is the right place for that tree for that time period. He will sustain it there, and will eventually bring it out of that place of chastisement. For He will not leave us in our unbelief, sinfulness, or harsh conditions, but will bring us to repentance and restoration.

One characteristic of the heath that ties into this point is its special need for human care. It is a cultivated plant that must be specifically planted by man, and maintained by man. If you want heath in your yard, you must plant it, for it will not grow up on its own.

In the same way, we as human beings must be both planted and maintained by the Lord. Without Him, we would not exist. And without Him, we could not continue to live.

This is true in normal circumstances, and how much more is it true in adverse conditions. There are times in life when the Lord must send us into adverse conditions because of our sin. He must send us, like the heath, to places where we are not naturally supposed to be. This could be to a hospital bed, though not every hospital visit is because of sin. It could be anywhere that we are not naturally designed to be. Yet the Lord, Who places us there for our good, sustains us Himself so that we may learn the lesson He has for us.

In the case of Moab in Jeremiah 48, this was done to save their lives. The Lord said, *"Flee, save your lives, and be like the heath in the wilderness."* *(Jeremiah 48:6)* They had been trusting in their works and treasures. To turn them from this false trust, the Lord had to plant them as a heath in the wilderness. There they clearly stood out, and it was well known that they didn't belong there. They were like someone walking into a casual place with a shirt and tie on. Clearly this person would be out of place. But this was what Moab had to endure in order to learn not to trust in its own works and treasures.

For the wayward men of Judah, the Lord seemed to be especially harsh. He said, *"For he shall be like the heath in the*

desert, and shall not see when good cometh." (Jeremiah 17:6)
What does the phrase mean, *"he . . . shall not see when good
cometh"*? Does this mean there is no hope of any deliverance
from this judgment? Is God's goal here not salvation and the
ultimate good of Judah, but total destruction?

For a proper understanding of this phrase, we again must
know something of the heath. The heath cannot take fertilizer.
Fertilizer, instead of helping it, will cause it to get sick and die. If
you must fertilize the heath, then do it only when you first plant
it. After it is planted, leave it alone, for any more fertilizer could
kill it.

The phrase concerning the men of Judah has to do with the
purpose of God's planting them in the desert. They were placed
there for judgment. If they were to begin to wither under that
judgment, they were not to be fertilized, though one might think
fertilizing them would be doing "good" to them. This would in
reality harm them and cause them to become more sick and weak.

Over-fertilizing any plant may make it not produce fruit,
which God desires that we all produce. Therefore fertilizer may
destroy them totally and thereby interfere with God's purpose of
ultimate good and restoration.

But even if fertilizer would help them, God still did not want
it. For He did not want them to become strong and powerful in a
strange land. They were placed there for judgment, not to grow
and prosper. Doing good to them would also defeat the purpose
of God's judgment, for they would not feel the full weight of it
and learn the lesson He had for them.

Instead man must not interfere with God's judgment. God
will sustain those He judges just as much as they need, but He
will also allow them to suffer the punishment they need in order
to learn their lesson. This does not mean we are not to pray for
those in judgment. We should. However, we should pray, not so
much for deliverance, but for God to help them endure until the
judgment is ended. Let us be thankful for the good God that we
serve, Who brings both punishment and sustenance in just the
right amounts and at just the right times, Who is working all
things together for our ultimate good.

25
Hemlock

Hemlock is found in the prophetic books of Hosea and Amos. Hosea says, *"They have spoken words, swearing falsely in making a covenant: thus judgment springeth up as hemlock in the furrows of the field." (Hosea 10:4)* The word used for hemlock here is the Hebrew word *rosh* (ראש), which is usually translated "gall," or simply "poison."

However, in Amos, a different Hebrew word is used. Amos says, *"Shall horses run upon the rock? will one plow there with oxen? for ye have turned judgment into gall, and the fruit of righteousness into hemlock." (Amos 6:12)* The word used here is *la'anah* (לענה), usually translated "wormwood."

Many have tried to say that hemlock is really gall, and some have said that it is really wormwood. But both gall and wormwood are mentioned elsewhere in the Bible specifically. If either one were the same as hemlock, it and hemlock would have been referred to consistently by just one name throughout the Bible.

Also gall and wormwood are different from each other. In at least four verses they are mentioned together, as in Deuteronomy 29:18, *"gall and wormwood,"* in which cases it is clear that they are separate plants.

A couple more points should make it clear that hemlock is not the same as gall. First, if it were, the verse in Amos would be redundant, saying, "ye have turned judgment into **gall**, and the fruit of righteousness into **gall**." Another point is that gall is simply a general term for bitterness, not necessarily a poison. It can be applied to bitter herbs, or more commonly, to a bitter drink.

Hemlock, on the other hand is not bitter, but is poisonous. Biblical hemlock is different from both gall and wormwood. It is real hemlock, but it is not to be confused with the hemlock of North America, which is not poisonous.

American hemlock is classified with the Latin name *Tsuga*. It is usually found on the west coast, the east coast, and the northeast. This American hemlock looks very similar to the fir.

The Biblical hemlock is classified as *Conium maculatum*. It is in the carrot family and even looks like a wild carrot. We have mentioned before that parsley is also in this family. But while the hemlock is highly poisonous, neither parsley nor carrots are poisonous.

This hemlock is not a large tree like the hemlocks of North America. Instead, it grows to be only about three or four feet high. It is more of a bush than a tree. But because it is a woody plant, it comes within the scope of our study.

We now examine in more detail the two Biblical references to the hemlock. Amos says, *"Ye have turned judgment into gall, and the fruit of righteousness into hemlock." (Amos 6:12)* Taking the characteristics of gall and hemlock, this verse can read, "Ye have turned judgment into bitterness (gall), and the fruit of righteousness into poison (hemlock)." This reading shows us a clear trail of spiritual destruction. First bitterness creeps into the heart and life. When it does, that bitterness causes the fruit of that life to become poisonous, and eventually this results in spiritual death. This is a sober and serious warning not to allow even a hint of bitterness within our hearts.

Hosea uses the hemlock to prophesy against the sinful nation of Israel, saying, *"They have spoken words, swearing falsely in making a covenant: thus judgment springeth up as hemlock in the furrows of the field." (Hosea 10:4)* First, in the natural realm, we notice where the hemlock grows, in the *"furrows of the field."* It is a well-known fact that hemlock needs lots of water to grow. For this reason, some have said that it couldn't have grown in Israel. But again, we must remember that the Israel of the Bible was much greener than the dry Israel of today. Still, even in the lush environment of Biblical Israel, the water-loving hemlock still needed the water of the low places in the field.

But the much more important lesson in this verse is its spiritual lesson. The key is the phrase *"swearing falsely."* Israel had been guilty of speaking, listening to, and believing lies. Lies may not seem like much at first. They may go unpunished for several years. But just as fruit takes a while to ripen, it also takes a while for poison to take effect.

Socrates recorded the slow, but steady, effect of hemlock when he took it to end his life. He wrote that his extremities began to go numb first. Then the numbness slowly worked its way up to his heart, and there the poison finally killed him.

Sin does the very same to us. If we believe a falsity, it may take time, but it will first cause us to become numb to sin. We will then allow more sin to enter in until, eventually, it kills us spiritually, and at times also physically. This is what happened to the nation of Israel. They swore falsely, and God's judgment sprang up upon them as hemlock, with its slow, but sure, killing effect.

But the Israelites are not the only ones guilty of sin and facing judgment. We in the United States are equally guilty. Chiang Kai-shek, the president of Nationalist China, once said, "Japan is a disease of the skin, but communism is a disease of the heart." He was right in principle. Today as Americans, we may look about our country and bemoan the state of our government and society. But this is merely a disease of the skin. The far deeper problem is sin. Sin is a disease of the heart.

This same principle is also found in tree diseases. A tree can get a disease on the outside, a "disease of the skin." This is a problem, but it is not the worst problem a tree can have. The worst problem is a disease called "heart rot." This disease can affect either the whole tree, or just some of its limbs. With this disease, the outside of the tree looks good, but the center, the very heart of the tree or limb, rots out.

On the outside, the tree looks full of leaves and free of problems. This is because most nutrients are carried up the tree within two inches of the bark. The rotted heart may go unnoticed for some time. But when a wind comes, it will easily blow the weakened tree over. Disease of the heart, though unseen, is deadly.

This principle can also be applied to the church today. Many churches today look good outwardly. Men are dressed in suits and ties, and ladies in dresses. There are stained glass windows, steeples, TV screens, and bands. Yet there is no substance on the inside. They are rotted on the inside. Thus these churches are weak, and they are on a course to sure destruction.

Let us beware of this danger in our lives. Let us not allow the hemlock of sin to enter our hearts. There it will work slowly to numb us and eventually to kill us. Let us instead guard our hearts that we might be living and vibrant souls, full of many years and useful life with which we can serve Christ.

26

Human Attributes Given to Trees

In the Bible, many times human attributes are ascribed to trees. The economy of Israel was largely based upon agriculture, so the Israelites would have been able to see many similarities and analogies between plant life and human life. In Isaiah, trees are said to clap. Some trees are said to have a heart. Elsewhere, to trees are ascribed the activities of listening, standing, or even communicating to one another.

We continue this practice today, calling tree branches arms or saying that potatoes have eyes, or even speaking of tree "wounds," which is a misnomer. We also turn the analogy around and call human arms and legs limbs or the human torso a trunk. Such analogies, because they are so clear to the human mind, are used throughout the Bible to teach important truths. We will examine them in this chapter, beginning with that of trees communicating to one another.

Trees Talking

In Judges 9, a parable of trees talking to one another is given to illustrate the political situation of Israel just after the death of Gideon. At first the thought of trees talking to one another seems odd and silly. But even in the natural realm, it is not. Trees can and do communicate with one another, though not audibly, just as people can communicate without words. Deaf people communicate through sign language with their hands. A wife can communicate to her husband with only a look or a nudge.

A study of this inaudible communication between trees was once made on willow trees. A certain group of willows was infected with a caterpillar that is harmful to willows. After this local infestation, willow trees three miles away were noted as having an increase in the chemical that repelled these caterpillars. How did trees three miles from the caterpillars know that they were in the vicinity? We don't know, but this does reveal that there is some type of communication taking place.

A similar study was made in Africa on the mupone tree. The kudu deer, which eats mupone trees, was released into an enclosed area containing mupone trees. When he began to eat these trees, other mupone trees not in the enclosed area increased in an ammonia chemical that tastes bad to the kudu deer and keeps him from eating the tree.

Plants communicate, not only to one another, but also to animals. The above studies are examples of this, in repelling the caterpillars and the kudu deer. Another example is the acacia tree. This tree puts out a nectar that attracts ants. The ants then in turn repel other insects. In this way, the acacia communicates a message of welcome to the ants, but one of hostility to other insects.

Plants and man even communicate with one another in some ways. Well-documented studies have been made on the effects of man's music upon plants. When plants are exposed to rock music, they become sick and some die. On the other hand, when they are immersed in classical music, they survive and some even thrive.

Plants communicate back to man in several ways. A rose calls man to come through its sweet smell. Other plants send forth offensive odors that warn man to keep his distance. Plants also wilt to communicate their need for water. While this is not audible speaking, all of these are instances of plant communication.

In Judges 9, the communication of trees is used to illustrate Abimelech's rise to become the ruler over Shechem. This account took place during the days of the judges. The kingdom had not yet been established. God Himself was King at this time, and He ruled Israel through the judges. Gideon had held this position of judge but had recently died. While he had lived, if he had wanted to become king, he could have done so, but instead he recognized God as King.

However, at Gideon's death, he left many sons, all of whom were rightful heirs to his position. Abimelech, the son of Gideon and one of his handmaids, saw in this an opportunity for power. He desired to reign as king over Shechem. Accordingly he hired men, and together they killed seventy of his brethren in an attempt to destroy all rivals. Then Abimelech was crowned king of Shechem.

Yet in his destruction of his brethren, he missed the youngest, Jotham, who had hidden himself. When Jotham then heard of the anointing of Abimelech, he came to Shechem and spoke this parable to the men of Shechem. *"The trees went forth on a time to anoint a king over them; and they said unto the olive tree, Reign thou over us." (Judges 9:8)* Jothan used the group of trees as a picture of the men of Shechem who wanted a king.

First they went to the olive and asked him to reign over them. *"But the olive tree said unto them, Should I leave my fatness, wherewith by me they honour God and man, and go to be promoted over the trees?" (Judges 9:9)* The olive pictures another son of Gideon, one of those whom Abimelech had killed. The olive tree is the most noble and useful of the trees listed here. It produces anointing oil, fruit, and numerous benefits to God and man. It certainly would have made a better king. But this son did not want a kingdom. The olive had no ambition to rule over others. He simply desired to fulfill his place of usefulness, to produce his own fruit.

Wise men, like this son no doubt was, choose to be useful rather than great. They choose humble positions like those of pastor, teacher, or missionary. These positions, though quiet and often unnoticed, are more influential and have more real use than the great positions of president or senator. People in the great positions often fail to be truly useful because power corrupts. Rather, the true place of greatness is in the place of the most usefulness.

The trees then turned to the fig tree, another son of Gideon. *"And the trees said to the fig tree, Come thou, and reign over us." (Judges 9:10)* But the fig tree's response was the same as the olive's. *"But the fig tree said unto them, Should I forsake my sweetness, and my good fruit, and go to be promoted over the trees?" (Judges 9:11)*

So the trees turned to the vine, yet another son. *"Then said the trees unto the vine, Come thou, and reign over us." (Judges 9:12)* But again their call was refused. *"And the vine said unto them, Should I leave my wine, which cheereth God and man, and go to be promoted over the trees?" (Judges 9:13)*

While the fig and vine did not have as wide a range of usefulness as the olive, they nevertheless yielded some benefits and would have made better kings than Abimelech. But all these sons recognized the importance of usefulness over greatness. They chose to remain in their place of usefulness rather than seeking power for themselves.

Yet there is one tree that was forgotten by the men of Shechem. They called upon the honorable olive, the sweet fig, and the cheering vine, but they forgot to call upon the chief tree of the Middle East. They forgot the cedar tree. In the Bible, the cedar tree is a picture of the head of Israel, the Davidic throne, the Messiah. As in Ezekiel 17 which we looked at in chapter 16, the cedar pictures Jesus Christ. The men of Shechem failed to look to their rightful King, the Lord Himself.

So they eventually turned to the bramble, Abimelech. *"Then said all the trees unto the bramble, Come thou, and reign over us." (Judges 9:14)* The bramble, hungry for power, was all too willing to accept the kingship. *"And the bramble said unto the trees, If in truth ye anoint me king over you, then come and put your trust in my shadow: and if not, let fire come out of the bramble, and devour the cedars of Lebanon." (Judges 9:15)* Abimelech's real hatred was for the cedars of Lebanon, the Messianic throne of Christ. It was the Lord Whom he viewed as his greatest rival. It was Jesus Christ Whom he sought to dethrone.

The bramble is the lowest of the trees. It has thorns, which scrape and irritate all about it. This is exactly what Abimelech would do when he obtained power. He would irritate the subjects under his dominion.

The book of Ecclesiastes speaks of the tendency of powerful positions to attract corrupt men. *"There is an evil which I have seen under the sun, as an error which proceedeth from the ruler: Folly is set in great dignity, and the rich sit in low place." (Ecclesiastes 10:5-6)* The general principle is that rulers are men of folly because they are often interested only in themselves. This was certainly true of Abimelech. Interested in himself, he ruled in folly.

Thus the bramble was not the answer. The olive would have been better. But even better yet would have been the cedar. Christ wants to be the supreme Ruler of men's lives. In order to reestablish His rightful dominion in Israel, He must therefore bring down Abimelech and remove his kingdom.

Trees Walking

Another place in the Bible where trees are likened to man is Mark 8. Here Christ was met by a blind man who asked Him to touch him. In response, Christ *"took the blind man by the hand, and led him out of the town; and when he had spit on his eyes, and put his hands upon him, he asked him if he saw ought."* *(Mark 8:23)*

The blind man then *"looked up, and said, I see men as trees, walking."* *(Mark 8:24)* How did the blind man know what a tree looked like? No doubt he had touched one before. He probably had conjured up an image in his mind of a tall, straight object with arms extending from it. When his eyes were touched and he saw men for the first time, the tree was the first object that came to his mind.

Eliphaz in the book of Job had a similar experience in which he could not discern the form of a spirit that passed by him. He said, *"Then a spirit passed before my face; the hair of my flesh stood up: It stood still, but I could not discern the form thereof: an image was before mine eyes, there was silence, and I heard a voice, saying, Shall mortal man be more just than God? shall a man be more pure than his maker?"* *(Job 4:15-17)* In times when a sight is entirely foreign to us, we may not be able to think of how to express it. This was the case with the blind man. Seeing a foreign sight, a man, he expressed it as a tree.

Christ then finished the work He had begun for the blind man. *"After that he put his hands again upon his eyes, and made him look up: and he was restored, and saw every man clearly."* *(Mark 8:25)*

Christ certainly could have healed the man fully at the first. But He chose instead to heal him partially, wait for his response, and then heal him fully. Perhaps He did this to illustrate the way He deals with us. He gives us first a little light. If we receive that light and respond to it, then He will give us more light.

An example of this is seen in the discipline of reading the Bible. Some people complain that they cannot understand the Bible, and they therefore cease reading it altogether. But this does not help their understanding. It gives them only more darkness. The only way to receive more light is to act upon the light that is already given, to obey what is understood and to continue to read and study the Bible because it is commanded. As we do this, the more we read and study, the more insight the Lord will give us.

Psalm 119:105 lays down this principle, *"Thy word is a lamp unto my feet, and a light unto my path." (Psalm 119:105)* The word used here for "lamp" is one that denotes a foot lamp. Like a foot lamp, the Bible gives just enough light for the next step. If we take that step, then the next step will be illuminated. But if we turn off the light completely, we won't know where we are going.

The second word used in this verse, the word "light," denotes very bright light, like the shining of the sun and the light of day. This word gives promise of greater light to come.

I Corinthians 13 also reminds us of our need for increased light. *"For now we see through a glass, darkly; but then face to face: now I know in part; but then shall I know even as also I am known." (I Corinthians 13:12)* Like the blind man whom Christ healed, we now see in part. Our sight is not yet whole. We see spiritual truths like heaven only in our hearts, not yet with our eyes. In this way we have internal sight, but not yet external sight. That external sight, that greater light, will come one day when we see Christ *"face to face."*

Trees Standing

In Ezekiel 31, the Lord likened Pharaoh, the king of Egypt, to a cedar tree. We have already come across this passage several times in our study of specific trees, and now we will look at in detail as it demonstrates human attributes given to trees. With the cedar tree, the Lord demonstrated these human attributes: first Pharaoh's ***great power***, then his ***great pride***, and finally his ***great fall***.

The Lord opened this prophecy by speaking to Ezekiel. *"And it came to pass in the eleventh year, in the third month, in the first day of the month, that the word of the LORD came unto me, saying, Son of man, speak unto Pharaoh king of Egypt, and to his multitude; Whom art thou like in thy greatness?"* (Ezekiel 31:1-2) Pharaoh of Egypt had exalted himself and wanted to be great like the king of Assyria.

The Lord went on to describe the loftiness of the king of Assyria. *"Behold, the Assyrian was a cedar in Lebanon with fair branches, and with a shadowing shroud, and of an high stature; and his top was among the thick boughs."* (Ezekiel 31:3) Like the king of Assyria, Pharaoh had a shadowing shroud. This symbolizes the extent of his empire. Many nations were under his shadow, relying upon him for dominion and support, and his empire encompassed a large territory. The thick boughs of this cedar illustrate prosperity and greatness. Such greatness was what the king of Egypt sought after, to flourish with thick boughs like a cedar of Lebanon.

It must be noted, however, that this is ***a*** cedar, not ***the*** cedar. This is not the Davidic and Messianic throne of Israel. Pharaoh wanted greatness and splendor ***like*** that of Messiah, but he could never be ***the*** Messiah.

In verse 4, Pharaoh was said to be the center to and from which water flowed. *"The waters made him great, the deep set him up on high with her rivers running round about his plants, and sent out her little rivers unto all the trees of the field."* (Ezekiel 31:4) Pharaoh watered and supported all the nations under his rule, making them, and thereby making him, prosperous. *"Therefore his height was exalted above all the trees of the field, and his boughs were multiplied, and his branches became long because of the multitude of waters, when he shot forth. All the fowls of heaven made their nests in his boughs, and under his branches did all the beasts of the field bring forth their young, and under his shadow dwelt all great nations. Thus was he fair in his greatness, in the length of his branches: for his root was by great waters." (Ezekiel 31:5-7)*

Verses 8 and 9 use the metaphor of Eden, the garden of God. *"The cedars in the garden of God could not hide him: the fir trees were not like his boughs, and the chesnut trees were not like his branches; nor any tree in the garden of God was like unto him in his beauty. I have made him fair by the multitude of his branches: so that all the trees of Eden, that were in the garden of God, envied him." (Ezekiel 31:8-9)*

This is not the Eden of Genesis, but is used as a description. The lush plain of Sodom had been similarly likened to Eden when Lot chose it. *"And Lot lifted up his eyes, and beheld all the plain of Jordan, that it was well watered every where, before the LORD destroyed Sodom and Gomorrah, even as the garden of the LORD, like the land of Egypt, as thou comest unto Zoar." (Genesis 13:10)*

Here in Ezekiel the term is used to picture the nations of the world as trees in a garden owned by the Lord. Of all the nations, Pharaoh was the largest tree. His greatness could not be compared to the other trees. His empire was large, and his branches spread his protective rule over even the greatest of the trees about him.

Some have suggested that the term Eden, or garden of God, is a figurative term for the great hanging gardens of Babylon, thus likening Pharaoh to Nebuchadnezzar. These hanging gardens were magnificent, said by one account to be 52 miles long, but it seems more likely that the term simply speaks of the nations as God's garden, just as Eden was the garden of God.

But even the ***great power*** of Pharaoh in the mist of God's garden was not enough to sustain him from God's wrath. He became lifted up with ***great pride,*** and therefore he suffered a ***great fall***. God said, *"Because thou hast lifted up thyself in height, and he hath shot up his top among the thick boughs, and his heart is lifted up in his height; I have therefore delivered him into the hand of the mighty one of the heathen; he shall surely deal with him: I have driven him out for his wickedness." (Ezekiel 31:10-11)*

Trees Prideful/Fainting/Mourning

After many trees had been broken and scattered by the great fall of Pharaoh, the remaining trees fainted. *"Thus saith the Lord GOD; In the day when he went down to the grave I caused a mourning: I covered the deep for him, and I restrained the floods thereof, and the great waters were stayed: and I caused Lebanon to mourn for him, and all the trees of the field fainted for him."* (Ezekiel 31:15)

Having lived so long under Pharaoh's shadow, the smaller nations had become dependent upon him. They now fainted and shook at his absence. *"I made the nations to shake at the sound of his fall, when I cast him down to hell with them that descend into the pit: and all the trees of Eden, the choice and best of Lebanon, all that drink water, shall be comforted in the nether parts of the earth."* Especially those nations that had been united with Pharaoh, drinking water with him, were affected.

Thus Pharaoh, by his pride, brought down himself and all the nations about him. *"They also went down into hell with him unto them that be slain with the sword; and they that were his arm, that dwelt under his shadow in the midst of the heathen. To whom art thou thus like in glory and in greatness among the trees of Eden? yet shalt thou be brought down with the trees of Eden unto the nether parts of the earth: thou shalt lie in the midst of the uncircumcised with them that be slain by the sword. This is Pharaoh and all his multitude, saith the Lord GOD."* (Ezekiel 31:16-18)

At the first reading, this seems like a tragedy, as trees fall and faint, and indeed it is a tragedy for the proud Pharaoh. However, a surprising benefit came to the fainting trees. When a large tree falls, it opens up the way for sunlight to reach the smaller trees. The very tree that the small nations thought was supporting them, was in reality suppressing them and hindering them from growing. Thus the fall of Pharaoh proved to be a benefit to the other nations, not a harm.

The same is true of all prideful men. Lifting up themselves, they suppress the light of the Gospel. They hinder that light from shining to others. Only those who do not eclipse the light with their own supposed greatness can receive the growth that it gives. And only when the prideful are brought down can that light fully shine into humble hearts.

But perhaps an even more sinister figure is alluded to here. A reference to the garden of Eden, such as "the trees of Eden" or "the garden of God," appears seven times in this chapter of Ezekiel, more than in any other chapter of the Bible. This chapter is about **_great power_**, **_great pride_**, and a **_great fall_**. Who else displayed these qualities but Satan himself, the tempter who lurked in the garden of Eden?

"Thou hast been in Eden the garden of God; every precious stone was thy covering, the sardius, topaz, and the diamond, the beryl, the onyx, and the jasper, the sapphire, the emerald, and the carbuncle, and gold: the workmanship of thy tabrets and of thy pipes was prepared in thee in the day that thou wast created. Thou art the anointed cherub that covereth; and I have set thee so: thou wast upon the holy mountain of God; thou hast walked up and down in the midst of the stones of fire. Thou wast perfect in thy ways from the day that thou wast created, till iniquity was found in thee. By the multitude of thy merchandise they have filled the midst of thee with violence, and thou hast sinned: therefore I will cast thee as profane out of the mountain of God: and I will destroy thee, O covering cherub, from the midst of the stones of fire. Thine heart was lifted up because of thy beauty, thou hast corrupted thy wisdom by reason of thy brightness: I will cast thee to the ground, I will lay thee before kings, that they may behold thee." (Ezekiel 28:13-17)

"For thou hast said in thine heart, I will ascend into heaven, I will exalt my throne above the stars of God: I will sit also upon the mount of the congregation, in the sides of the north: I will ascend above the heights of the clouds; I will be like the most High." (Isaiah 14:13-14)

"Thou hast defiled thy sanctuaries by the multitude of thine iniquities, by the iniquity of thy traffick; therefore will I bring forth a fire from the midst of thee, it shall devour thee, and I will bring thee to ashes upon the earth in the sight of all them that behold thee. All they that know thee among the people shall be

151

astonished at thee: thou shalt be a terror, and never shalt thou be any more." (Ezekiel 28:18-19)

Trees Clapping

In Isaiah, another human attribute is given to trees. This is the action of clapping. Isaiah says, *"For ye shall go out with joy, and be led forth with peace: the mountains and the hills shall break forth before you into singing, and all the trees of the field shall clap their hands." (Isaiah 55:12)*

While trees do not have hands like we think of them, yet we have probably all heard them "clap." Both the cottonwood and the aspen are capable of producing this sound. This is because of the shape of their petiole, or leaf stem. Most trees have a round petiole, but the cottonwood and aspen have a flat one. Thus when the wind blows, their leaves bend back and forth, hitting one another. In a large tree, this activity sounds like clapping. And in a whole forest of these trees, the sound can be very loud.

This ability of the cottonwood and the aspen also gives them an advantage in wind. Because of their flat stems and weak wood, they can sway in the wind without breaking. This makes them more hardy and able to endure storms.

This verse in Isaiah refers to the return of Israel after the Babylonian captivity. The clapping of the trees pictures the rejoicing that accompanied Israel as she returned home. As she passed through other nations, they came out and stood on the mountains or among the trees. There they sang and clapped their hands to join with Israel in her rejoicing.

Even the land itself seemed to rejoice at the return of Israel. *"Instead of the thorn shall come up the fir tree, and instead of the brier shall come up the myrtle tree: and it shall be to the LORD for a name, for an everlasting sign that shall not be cut off."* *(Isaiah 55:13)* Instead of finding thorns and briers in their land, Israel found the fir and the myrtle.

This prophecy also has a double fulfillment in the return of Christ, the final restoration of Israel. The curse of thorns and briers, which has been since the fall, will be no more. Symbolically also, the old nations and the old rule will be gone.

The pleasant fir and myrtle will be in their place. Christ will reign supreme, and the joy will be complete.

Trees Rejoicing, Talking, and Howling

Another place where trees are said to act with human emotions and responses is Isaiah 14. *"Yea, the fir trees rejoice at thee, and the cedars of Lebanon, saying, Since thou art laid down, no feller is come up against us." (Isaiah 14:8)* In this passage, the destruction of the king of Babylon by the Medes has just been foretold. This verse then prophesies the response of the trees, saying first that the fir trees will rejoice. The oppression of Babylon will finally be removed, and this will give these trees great cause for rejoicing.

The verse goes on to prophesy the response of the cedars, that they will say in triumph, *"Since thou art laid down, no feller is come up against us."* Israel, the cedars in this case, is saying that the king of Babylon has laid down his axe with which he had cut down Israel. Now there is none to lift up an axe against Israel and fell these cedars of Lebanon. Together the firs and cedars will rejoice over the destruction of Babylon.

Yet contrast this response of rejoicing with the response of the fir tree in Zechariah. *"Howl, fir tree; for the cedar is fallen; because the mighty are spoiled: howl, O ye oaks of Bashan; for the forest of the vintage is come down." (Zechariah 11:2)* Here the fir tree howls, for the great cedar is fallen.

Again the cedar represents Israel, or the temple which was physically made of cedar. Israel and the temple are said to have fallen because they have rejected Christ. The fulfillment of this prophecy is found at the very opening of Christ's earthly ministry in the book of John. *"He came unto his own, and his own received him not." (John 1:11)*

Because of this rejection of Christ, Israel fell. Israel had given to the world the very Scripture that prophesied of the Messiah. She had been the nation that held up the light of the true God to all the others. Yet when Messiah came, she herself rejected Him. Thus the fir tree and even the oaks of Bashan howled over her fall. The great cedar of Lebanon to which all the trees had looked was now fallen.

We today also howl when one of our great preachers or ministers falls into sin or rejects the Biblical teachings he once preached. Indeed, this is a tragedy. Yet let us take heed. If these learned leaders, who have studied the Scriptures, can fall, how much easier is it for the rest of us to fall?

One final note of comfort is that this falling is not total destruction. Israel has fallen, but is not destroyed. If she will turn to her Messiah, God yet gives us hope that the cedar will grow again. He also gives us hope as individuals, that though we may fall, if we will repent, we also will grow again. Job said, *"For there is hope of a tree, if it be cut down, that it will sprout again, and that the tender branch thereof will not cease." (Job 14:7)* This is the hope to which we cling, that God will allow us to learn from our fall, repent and return to truth, and grow again to be a profitable and useful tree.

27
The Hyssop

The hyssop is a very well known Biblical plant, rich in Biblical symbolism. Its Hebrew name is *ezowb* (עֵזוֹב), which even sounds like the English word "hyssop." In seeking to identify it, many plants have been suggested. Two unlikely suggestions are rosemary and mint. More plausible is the caper, called by the Latin term *Capparis spinosa*. Still more likely, I believe, is *origanum*, or marjoram, of which oregano is one species. At times, the Biblical hyssop looks more like the caper, and at times more like *origanum*, but based on the reference that we will look at in I Kings, I believe it is *origanum*.

The hyssop is a short plant, growing to be only about three or four feet tall. It is not as small as most flowers, but neither is it as tall as most trees. And so it does not fall neatly in any category. But it is a woody plant, and so we will examine it here. It is of the gymnospermia order, which means that its seeds are on the outside of its fruit, like a strawberry. This characteristic is also called "naked seed," from the Greek *gymnos* (naked) and *sperma* (seed).

The hyssop has an unusual pattern of growth. It grows in three stalks called bunches. Each of these three bunches also has three branches. In these multiples of three in which three branches make up one bunch, many see a picture of the Trinity, one God made up of three persons.

The hyssop is first mentioned in Exodus. Here the Lord was preparing to send the death angel upon the land of Egypt to kill every firstborn son. But before He sent the plague, He provided a way of deliverance. This He communicated, not to the common people, but to the elders, the leaders of Israel. This will be important later on as we continue to study the hyssop. *"Then Moses called for all the elders of Israel, and said unto them, Draw out and take you a lamb according to your families, and kill the passover. And ye shall take a bunch of hyssop, and dip it in the blood that is in the bason, and strike the lintel and the two side posts with the blood that is in the bason; and none of you shall go out at the door of his house until the morning." (Exodus 12:21-22)*

The hyssop here was used to apply the blood of the lamb to the lintel and the two side posts of the door. It is ideal for this task due to its structure. It has feather-like needles and leaves, the ends of which are soft like cotton balls. These feathery leaves and the "cotton balls," spread out along the stem, allow the hyssop to work like a paintbrush, picking up, holding, and applying liquid well.

But the most important thing to notice in this account is the blood. The hyssop was not the important thing, but the blood it carried. Also, nationality was not important. The death angel made no distinction between Egypt and Israel. He was looking for one thing and one thing alone, the blood. At that sight only, he would "pass over," and the life of the son would be saved. As the beloved hymn says, "When I see the blood, I will pass over you." Indeed when he saw the blood, he passed over.

The same is still true today. It doesn't matter what nationality we are, or what other distinguishing characteristics we may have. It is only the blood that Christ is looking for. Only that will give us eternal salvation. And it is available to all, no matter who we are.

The hyssop is also mentioned in I Kings 4. This passage is describing the wisdom and knowledge of King Solomon, and it says of him, *"And he spake of trees, from the cedar tree that is in Lebanon even unto the hyssop that springeth out of the wall: he spake also of beasts, and of fowl, and of creeping things, and of fishes." (I Kings 4:33)* Solomon had extensive knowledge on many subjects, plants, animals, gardening, music. He knew a lot about a lot of things. But instead of being the "jack of all trades and master of none," he was truly the master of all the fields he took up. The hyssop is used here to show his mastery of the field of botany, to show that his knowledge covered from one of the tallest plants, the cedar, to one of the smallest plants, the hyssop.

This teaches us something of importance in God's eyes. Though the hyssop is small, it is still important and worthy of mention. The cedar's height does not automatically make it more important. Instead, both the cedar and the hyssop are important, for both have their own uses for which they are specially designed. Neither could fulfill the other's purpose.

The hyssop here is described as *"the hyssop that springeth out of the wall."* The ability to grow out of a wall shows that the hyssop can grow in very austere conditions, in hot, dry, and low-soil areas. It is because of this description that I believe the Biblical hyssop to be *origanum*, not caper. *Origanum* grows out of rocky crevices and rock walls, whereas the caper does not.

Another place where the hyssop is mentioned is Leviticus 14. Here its qualities as a brush were used for purifying leprosy, both in one's flesh and in one's house. The early part of the chapter speaks of leprosy in the flesh, for which God commanded, *"And the priest shall go forth out of the camp; and the priest shall look, and, behold, if the plague of leprosy be healed in the leper; Then shall the priest command to take for him that is to be cleansed two birds alive and clean, and cedar wood, and scarlet, and hyssop."* (Leviticus 14:3-4)

Again we see that this command was not given to the common people. It was instead given to the priest, just as the command to apply the blood of the Passover had been given to the elders of Israel.

The priest was to apply the hyssop in this way. *"And the priest shall command that one of the birds be killed in an earthen vessel over running water: As for the living bird, he shall take it, and the cedar wood, and the scarlet, and the hyssop, and shall dip them and the living bird in the blood of the bird that was killed over the running water: And he shall sprinkle upon him that is to be cleansed from the leprosy seven times, and shall pronounce him clean, and shall let the living bird loose into the open field. (Leviticus 14:6-7)*

The process for purifying a house infected with leprosy is described later in the same chapter in verses 49-53. It is essentially the same, purifying the house with the blood of a bird, cedar wood, scarlet, and hyssop.

This process shows clearly the function of hyssop as a purging and purifying instrument. Not only does its brush-like quality apply something clean, such as blood, but it also carries away the unclean, such as leprosy. In this case of leprosy, this function is not merely symbolic. For even in natural terms, hyssop has anti-bacterial and anti-microbial qualities, and can be used to remove contamination.

However, purification from sin, which is the next use of the hyssop in the Bible, cannot be accomplished naturally. Here in this reference, Numbers 19, the purifying quality of hyssop was clearly miraculous.

Here a red heifer without blemish was slain and burned. While her body was burning, the priest was to *"take cedar wood, and hyssop, and scarlet, and cast it into the midst of the burning of the heifer." (Numbers 19:6)* Once the burning was finished, these ashes, including those of the cedar wood, hyssop, and scarlet, were to be collected up and mixed with clean water. This mixture was then a purifying and cleansing substance called a "water of separation." God said of it, *"It shall be kept for the congregation of the children of Israel for a water of separation: it is a purification for sin." (Numbers 19:9)*

The method by which this water cleansed involved the hyssop again, for this water of separation was to be applied with hyssop. *"And a clean person shall take hyssop, and dip it in the water, and sprinkle it upon the tent, and upon all the vessels, and upon the persons." (Numbers 19:18)* Also again we see that the person applying the hyssop was not an ordinary person, but had to be *"a clean person."*

Psalm 51, the next mention of hyssop, draws upon this purifying function and symbolism of hyssop. In this Psalm, David pleads, *"Purge me with hyssop, and I shall be clean: wash me, and I shall be whiter than snow." (Psalm 51:7)*

The One Who uses the hyssop here is Christ Himself. Again, like the elders and the priest and the clean person, the One Who applies the hyssop is not a common man, but One set apart.

The purging that Christ does here produces one that is *"whiter than snow."* This extreme whiteness reveals the thoroughness of the purging process. As we saw with the cleansing of leprosy, to purge something means to remove something from it. We can speak of purging food, removing impurities from it. We can also speak of purging an individual, removing impurities from his life. This is what is done here in Psalm 51.

This purging of sin can be illustrated with painting a wall. If a wall is blue and we want to paint it white, there are two ways we could do it. We could simply paint over the blue with white paint. But that wouldn't be purging. That would be grafting, overwriting what was already there, like overwriting a wall with graffiti. In fact, both "graft" and "graffiti" come from the same root, the Greek *graphein* (γραφειν), which means to write or draw.

The other way we could paint the blue wall white would be to take the blue paint off first and then paint the wall with white paint. This would be true purging. It would not be merely covering up what was already there, but truly purging it, removing it. This kind of painting will produce the brightest white wall. No blue will show through. It will be white like snow.

This is what Christ does with our sin. He doesn't merely cover it up, but He totally removes it. He removes the unclean and replaces it with Himself, the clean. He purges us with hyssop and makes us *"whiter than snow."*

Another mention of the hyssop plant is found in Hebrews 9. While this is a New Testament reference, it refers back to Moses in the Old Testament and to purification with hyssop such as we saw in Leviticus and Numbers. *"For when Moses had spoken every precept to all the people according to the law, he took the blood of calves and of goats, with water, and scarlet wool, and hyssop, and sprinkled both the book, and all the people." (Hebrews 9:19)*

Moses here applied the blood and hyssop to the book and the people, purging and purifying them. Again the one applying the hyssop was not a common man. It was Moses, the very leader of Israel, just as in other references it has been the elders, the priests, the clean person, and even Christ Himself.

Here, as in both Leviticus and Numbers, we find blood, water, scarlet, and hyssop. The only thing missing here that is in the other references is cedar wood. I believe it is because of the nature of the cedar wood that it is left out here in the New Testament.

We covered the cedar tree in chapter 9, and there we saw that cedar has a preserving and anti-decay quality. In ancient Egypt, cedar wood was ground up and put into the coffins of pharaohs. This was done to keep insects away and to preserve the bodies.

This was probably also the symbolic reason for the cedar being used in the cleansing of leprosy. As a preserving and anti-decay agent, it was to symbolize preservation of the person or the house from any return of leprosy. It was to sustain their health after their cleansing and to prevent them from ever getting the disease again.

But in the New Testament, we no longer need the symbolic preservation of the cedar. It is Christ Himself Who does the preserving now. He is the preserving power that preserves both His Word, *"the book,"* and Christians, *"the people."*

The final reference to hyssop that we will examine is perhaps the best known. It is that found in John 19 at the crucifixion of Christ. *"After this, Jesus knowing that all things were now accomplished, that the scripture might be fulfilled, saith, I thirst. Now there was set a vessel full of vinegar: and they filled a spunge with vinegar, and put it upon hyssop, and put it to his mouth."* (John 19:28-29)

Here a sponge was placed upon hyssop. This begs for an explanation. With the absorbent quality of the hyssop, why was a sponge needed? Why not use the hyssop alone?

Perhaps the reason for this was that the hyssop is not very tall. Perhaps to reach up to the top of the cross, an extra length was needed. On the other hand, this passage may indicate that the cross was not as high as some have thought. If we say the Roman soldier could raise his hand to a height of eight feet, which is reasonable, and then add three or four feet on to that for the hyssop, we reach a height of only eleven or twelve feet for the cross.

This may be more accurate than the very high cross that is sometimes portrayed. Other Biblical references seem to indicate that low executions were the norm in Bible times. For example, when seven of Saul's descendants were hanged in II Samuel 21, it was necessary to keep animals away from them. *"And Rizpah the daughter of Aiah took sackcloth, and spread it for her upon the rock, from the beginning of harvest until water dropped upon them out of heaven, and suffered neither the birds of the air to rest on them by day, nor the beasts of the field by night." (II Samuel 21:10)* Had the bodies been hanged high above the ground, there would have been no danger of the *"beasts of the field"* molesting them. But since a constant guard was required, they must have been within reach of the beasts.

One exception is the gallows that Haman made for Mordecai. This gallows was *"fifty cubits high." (Esther 7:9)* But the emphasis placed on its height, repeated throughout the book of Esther, reveals that it was unusually high. This was not the normal height. Also hanging requires more height than crucifixion.

But perhaps a deeper reason is the symbolism of the hyssop. Recall who applied it in the previous passages. It was elders, priests, clean men, Moses, and Christ. And it was for the purpose of purification. But in this case, an unclean man was applying the hyssop. And he was applying it to Jesus Christ, the clean and pure, Who needed no purification.

Were the hyssop to have gone directly from the unclean Roman soldier to the clean and undefiled Son of God, this would not have been a fitting or proper picture. Jesus Christ needed no purifying. He had no sin that needed to be purged. Also an unclean man could not administer purification. And so in God's providence, a sponge was placed in between the hyssop and Christ, showing that He was already pure and needed no purification from unclean man. The hyssop here was not applied for purification, but was used as a stick to lift the sponge to relieve Christ's thirst.

A final beautiful picture is that the hyssop, when used to purify in the Old Testament, was always accompanied with water and blood. These were the very substances that flowed from Christ on the cross. Truly we can be thankful for the purification that is offered to us in Christ. His blood and water are applied to us through the hyssop, which He applies Himself. In this way only, our sins are purged, totally removed, and His purity is imparted in their place.

28
The Juniper Tree

The Biblical juniper tree is not the same as our North American juniper. As we will see when we look at I Kings, it was large enough for Elijah to sit under. This would be very difficult to do with our juniper tree. Ours is a small, ornamental, pyramidal evergreen, and doesn't provide much shade at all. It has the Latin name *Juniperus* and is similar to the redcedar. But the Biblical juniper is much different.

The Biblical juniper has the Hebrew name *rethem* or *rothem* (רתם). Both the heath and the fir have been suggested for its identity. The very small heath can be ruled out quickly, for it, like our juniper, would also be impossible to sit under. The fir can also be ruled out, for it is mentioned elsewhere in the Bible, and it is also not in the same family of trees as the juniper.

The key to identifying the Biblical juniper is to look at the meaning of its name in Hebrew. This name, *rethem* (רתם), means "broom." Knowing this, it is now easy to identify the Biblical juniper as a broom tree. It has small broom-like needles and arching branches, ideal for sitting under. It also flourishes in the deserts of southern Israel, which is where Elijah was when he sat under the juniper.

This Hebrew name is preserved in the Bible in Numbers 33, the first mention of the juniper tree. Describing the journeys of the children of Israel, this passage says, *"And they departed from Hazeroth, and pitched in Rithmah." (Numbers 33:18)* Rithmah is the feminine form of *rethem*. Thus to a Hebrew reader, the latter part of this verse would read that they "pitched in the land of broom trees." The picture is clear. In the hot desert, they found a place full of broom trees. There they found rest and refreshment sitting under these shady broom trees.

The juniper tree is also mentioned in Psalm 120. This Psalm asks the question, *"What shall be given unto thee? or what shall be done unto thee, thou false tongue?" (Psalm 120:3)* The answer comes with this heavy judgment upon the false tongue,

"Sharp arrows of the mighty, with coals of juniper." (Psalm 120:4)

This is a very severe judgment. The wood of the Mediterranean broom tree, or Biblical juniper, when it is burned, generates the hottest heat of most any tree in the Mediterranean area. It is still used by Arabs today when they want a very hot fire. When this wood is burned to make coals, the coals are also the hottest-burning coals. These hot coals were what God used to bring a fitting judgment upon the false tongue. Such a tongue was to be pierced with sharp arrows and burned with these extremely hot coals. Such a serious judgment reveals that God considers a false tongue to be a very serious problem. It is not to be taken lightly, but very seriously.

With this hot burning being a characteristic of the juniper, it is possible that it provided the coals used in the fiery furnace into which Shadrach, Meshach, and Abednego were cast. This fire was extraordinarily hot, seven times hotter than normal. The Bible doesn't say that juniper coals were used here, so we can't say they were for sure. But since they were the hottest coals known to man in the area, it is very likely that they were.

Another mention of the juniper is one to which we have already alluded, in which Elijah sat under a juniper. After Elijah's victory on Mt. Carmel and his slaying of the prophets of Baal, an angry Jezebel set out swiftly to kill him. At this threat, an unexpected thing happened to Elijah. He who had just called down fire from heaven before a whole nation of hostile Baal worshippers, suddenly became afraid of one woman.

I Kings 19 records the story, *"And when he saw that, he arose, and went for his life, and came to Beersheba, which belongeth to Judah, and left his servant there. But he himself went a day's journey into the wilderness, and came and sat down under a juniper tree: and he requested for himself that he might die; and said, It is enough; now, O LORD, take away my life; for I am not better than my fathers. And as he lay and slept under a juniper tree, behold, then an angel touched him, and said unto him, Arise and eat." (I Kings 19:3-5)*

Again, in this passage we see the nature of the juniper tree by Elijah's use of it. It is large enough to sit under and has arched branches, perfect for taking refuge under and finding rest in the hot desert.

Job also mentions the juniper tree when he is talking to his three friends, or more properly psychiatrists. They certainly seem more like psychiatrists than friends. It is of them that Job is speaking when he describes in chapter 30 the difference between his former companions and his present companions. He used to know men of high degree, like kings and princes. But everyone he used to know has left him. The only ones left about him are those of low degree, men who identify with his boils and his new poverty. The difference is that Job has been brought to this condition unwillingly, whereas his companions have gotten there through their own slothfulness and laziness.

Job describes his new companions as those *"who cut up mallows by the bushes, and juniper roots for their meat." (Job 30:4)* Anyone who would stoop to eat juniper roots must be desperate for food. Juniper roots are noxious, slightly poisonous. Only a starving man would be desperate enough to eat them. For if you get hungry enough, you'll eat almost anything. These men, Job's new companions, were lazy and wouldn't plant their own food. For this reason, they were starving and were forced to eat whatever they could find, even juniper roots. It is these lazy men that Job must now associate with because everyone else has left him.

In this brief chapter, we have seen that the juniper is a very interesting tree with quite a variety of Biblical uses. In it, we find much to instruct us. It reminds us of the seriousness of a false tongue. It warns against unnecessary fears such as Elijah's fear of Jezebel. And it shows the ugly consequences of sloth and laziness. Let us take these warnings to heart that this broom tree gives to us. Let us find rest under its shady branches by avoiding the dangers of which it warns us.

29
The Mulberry Tree

The mulberry tree appears in both II Samuel 5 and I Chronicles 14 when the Philistines came against David. Since these two accounts have almost exactly the same wording, we will look at only one, II Samuel 5. Here the Lord commanded David, *"And let it be, when thou hearest the sound of a going in the tops of the mulberry trees, that then thou shalt bestir thyself: for then shall the LORD go out before thee, to smite the host of the Philistines." (II Samuel 5:24)*

There are many varied suggestions for the identity of this tree. One is the acacia, but it is mentioned elsewhere in the Bible as shittim.

Another suggestion is the poplar or aspen. Both of these are in the Latin genus *Populus*, from which we get the name "poplar." These trees are suggested because of the clapping sound of their leaves. In the passage in II Samuel, David was to listen for a sound in the tops of the mulberry trees. The poplar and aspen both have a flat leaf stem which causes their leaves to shake in the wind and make a clapping sound. However, the Lord can make a sound in the top of any tree whenever He wishes, whether the leaf stem is conducive to that or not. All trees make some sound when wind blows through them. Also the poplar is already mentioned elsewhere in the Bible.

Another suggestion is the sycamore, which is also mentioned elsewhere. It is interesting that the sycamore's leaves are very similar to mulberry leaves, and that another name for the sycamore is "fig mulberry," or in Latin *Ficus sycamorus*. But even this is not enough to make it the Biblical mulberry in light of the Hebrew word for mulberry that we will shortly examine. The sycamore is interesting in its own right, and there is much we have to study when we come to it later in this book.

Before I began this study, I had originally thought that the Biblical mulberry tree was the black mulberry, which is related to the red mulberry that we have in North America, both being in the *Morus* genus. But as I began to study it, I discounted that and began to think the Biblical mulberry was a pear tree instead. But again as I studied that, I found that there is no evidence for it, except that the Septuagint translates it as such.

The Hebrew word for mulberry is the word *baca* (בכה), and it literally means weeping. This word is found transliterated in Psalm 84, which says, *"Blessed is the man whose strength is in thee; in whose heart are the ways of them. Who passing through the valley of Baca make it a well; the rain also filleth the pools."* *(Psalm 84:5-6)* This passage speaks of the valley of Baca, or literally, the valley of "weeping." It shows that even as we pass through this sorrowful valley, the Lord is with us to sustain us with abundant rain. This picture of God's presence in a valley of sorrow is reminiscent of Psalm 23, which says, *"Yea, though I walk through the valley of the shadow of death, I will fear no evil: for thou art with me; thy rod and thy staff they comfort me."* *(Psalm 23:4)*

The Hebrew word *baca*, or weeping, translates into English as "balsam." However this is not what many of us think of as balsam wood, the soft wood that is used to carve model airplanes. Though this soft wood is commonly called "balsam," its real name is "balsa," without the "m." It is called balsa for its softness, but it really comes from a linden tree.

The true use of the word "balsam," though, is as an adjective that means "weeping," used to describe a type of tree. For example, the balsam fir is a tree used as a Christmas tree. It is called balsam because it is a "weeping" fir, a fir that weeps sap. When this fir's leaves are pulled off, it weeps a white or silvery sap. While all firs give forth some sap, the balsam fir is named "balsam" because it weeps more sap than any other fir. Its cones are covered with this sticky, silvery sap, as are its branches and needles. Instead of calling this tree the balsam fir, if we use the literal translation, we could just as well call this tree the "weeping fir."

So to conclude our search for the identity of the Biblical mulberry, we take simply its Hebrew name, *baca*, translated, balsam. The Bible leaves it there and doesn't specify exactly what kind of balsam tree it is, and so we leave it there too. The

Biblical mulberry is a balsam, full of gummy resin which it freely gives forth.

We now turn to examine II Samuel 5, the passage on the mulberry tree, in detail. In the beginning of this account, David had just become king of Israel, and the Philistines had come out to test the new king. *"The Philistines also came and spread themselves in the valley of Rephaim." (II Samuel 5:18)*

In response to this threat, David did the right thing. He turned to the Lord. *"And David enquired of the LORD, saying, Shall I go up to the Philistines? wilt thou deliver them into mine hand? And the LORD said unto David, Go up: for I will doubtless deliver the Philistines into thine hand." (II Samuel 5:19)*

Just as the Lord had promised, David gained the victory. *"And David came to Baalperazim, and David smote them there, and said, The LORD hath broken forth upon mine enemies before me, as the breach of waters. Therefore he called the name of that place Baalperazim. And there they left their images, and David and his men burned them." (II Samuel 5:20-21)*

However, the Philistines came a second time. *"And the Philistines came up yet again, and spread themselves in the valley of Rephaim." (II Samuel 5:22)* This time the story was a little different. David still turned to the Lord, but this time the Lord had a different answer for him. *"And when David enquired of the LORD, he said, Thou shalt not go up; but fetch a compass behind them, and come upon them over against the mulberry trees. And let it be, when thou hearest the sound of a going in the tops of the mulberry trees, that then thou shalt bestir thyself: for then shall the LORD go out before thee, to smite the host of the Philistines." (II Samuel 5:23-24)*

This time David was not to go up immediately, but he was to wait patiently behind the Philistines. To David's credit, He obeyed this more difficult command, to wait. And again the victory was his. *"And David did so, as the LORD had commanded him; and smote the Philistines from Geba until thou come to Gazer." (II Samuel 5:25)*

In our lives as Christians, God often works like He did here for David. When we are new Christians, as David was at first a new king, God often gives immediate answers so that we see His power and His ability to sustain us. He does this so that we learn early to trust Him alone for all things. But as we grow, He has new lessons to teach us. He is no longer teaching just simple trust, but harder virtues to learn, such as patience.

For our own good, we must learn this lesson of how God deals with His children. Early on in our Christian lives, He may solve many problems for us. He may tell us immediately, "Go up," as He told David. But as we grow, we should not just expect everything to be done for us. God may at times tell us to wait, as He also did for David. This is to teach us patience, a difficult lesson to learn. Waiting takes more faith than going. But it is for our good.

An important key to note is that the Lord went before David both times. The first time, He said, *"Go up: for I will doubtless deliver the Philistines into thine hand." (II Samuel 5:19)* The second time, He said, *"Then thou shalt bestir thyself: for then shall the LORD go out before thee, to smite the host of the Philistines." (II Samuel 5:24)* In both cases, it was the Lord Who went before David.

In the first battle, the Lord was there, but not noticeably. David could easily have thought that he had won in his own strength. In fact, he even said in the I Chronicles account, *"God hath broken in upon mine enemies by mine hand." (I Chronicles 14:11)* The phrase *"by mine hand"* shows that David saw himself as one of the key actors in the battle. But the second time, it was clear that the Lord was the prime and only actor. David had no doubt that the battle was won in the Lord's strength, not his own. This encouraged his heart, not only then, but for all future battles as well.

In the same way, the lesson of the mulberry tree should be a great encouragement to our hearts, to know that the Lord has also gone before us. For all of our problems, both known and unknown, the Lord has already gone ahead of us to give us victory when we arrive.

30
The Mustard Tree

The mustard tree does not exactly fall into the category of woody trees which is our focus for this book. The mustard is not woody, and it is not a true tree by our standards of classification. But the Bible calls it a tree in both Matthew and Luke. So we also will call it a tree and include it in this study.

The Greek word for mustard is *sinapi* (σιναπι). It first appears in Christ's parables, and so we will begin there. When studying parables, we must always keep in mind their purpose. A parable is not necessarily truth itself, as in a true historical account, but it instead teaches a truth.

Matthew records this parable about the mustard. *"The kingdom of heaven is like to a grain of mustard seed, which a man took, and sowed in his field: Which indeed is the least of all seeds: but when it is grown, it is the greatest among herbs, and becometh a tree, so that the birds of the air come and lodge in the branches thereof." (Matthew 13:31-32)*

Mark records the same parable with slight variation. *"Whereunto shall we liken the kingdom of God? or with what comparison shall we compare it? It is like a grain of mustard seed, which, when it is sown in the earth, is less than all the seeds that be in the earth: But when it is sown, it groweth up, and becometh greater than all herbs, and shooteth out great branches; so that the fowls of the air may lodge under the shadow of it." (Mark 4:30-32)*

We notice that Mark does not call the mustard a tree like Matthew does, but an herb. Mark also does not use Matthew's term "kingdom of heaven," but rather "kingdom of God."

Luke records the same parable, calling the mustard a tree like Matthew, but using the term "kingdom of God" like Mark. *"Unto what is the kingdom of God like? and whereunto shall I resemble it? It is like a grain of mustard seed, which a man took, and cast into his garden; and it grew, and waxed a great tree; and the fowls of the air lodged in the branches of it." (Luke 13:18-19)*

After Christ gave this parable, He then went on to give another one very similar to it using the analogy of leaven instead of a mustard seed. The principle is this, that a small amount becomes great.

The mustard seed was the smallest seed a farmer in Biblical times would have planted. Yet it would grow into a very large plant. Indeed, it is understandable why the Bible calls the mustard a tree. Though it is not woody, it can reach ten feet in height, and a few have even grown to twelve feet. Its branches can reach about one inch in diameter, large enough for a bird to build a small nest on them. And so the mustard tree illustrates well how something very small can become very large.

This truth can be applied to both the Jewish nation and to Christians. Both began small, but have grown. The Jewish nation began with one man, Abraham, and even he had no children. But the book of Hebrews records, *"Therefore sprang there even of one, and him as good as dead, so many as the stars of the sky in multitude, and as the sand which is by the sea shore innumerable." (Hebrews 11:12)*

Likewise Christianity began small. It began with just a handful of disciples. It is the narrow way, the way with only a few travelers. But it is not supposed to stay small. Christ said, *"ye shall be witnesses unto me both in Jerusalem, and in all Judaea, and in Samaria, and unto the uttermost part of the earth." (Acts 1:8)* Christianity is supposed to grow and spread into the uttermost part of the earth.

An important key to this growth is to realize that it transcends us. It doesn't depend on us, and it doesn't all have to be accomplished in our lifetime. We don't have to go out all at once and conquer the world, or change our nation or town, or change anything in our own strength. It is not to be done through our efforts or talents or abilities. Rather it is to be done, and can only be done, through Christ. We merely plant the seed of Christ. Christ does the growing. Though the seed may seem small to us at first, we can be assured that Christ will grow it.

We do also have a responsibility for personal growth. Just as we grow physically and mentally, so we should grow spiritually as well. But even this growth is not done in our strength, but through Christ. All growth is up to the Lord.

An interesting fact about the mustard tree is that in some places of the world it is considered it a weed. In these places, it grows so prolifically that is considered obnoxious, and people want to get rid of it. In the same way, the world looks at Christians as a weed, something they want to eliminate. Christians are a constant reminder to the world of their sin. This is a problem to the world. They don't want this conviction. They want to go on their wrong way and feel good about it. And so they seek to eliminate Christianity.

But another fact about the mustard tree is that it is very difficult to get rid of it. Though many consider it a weed and seek to eliminate it, this is very hard to do. The same is true with Christianity. Only, it is not only hard to eliminate, but impossible. The world thinks it can get rid of Christianity and has tried for centuries and still tries to this day. But it never will, for it never can. Christ has promised that the gates of Hell shall not prevail against His church.

Another occurrence of the mustard tree in the Bible is in Matthew 17. Here Christ had just cast out a devil that the disciples could not cast out. When they came to Him and asked why they could not, this was His answer, *"Because of your unbelief: for verily I say unto you, If ye have faith as a grain of mustard seed, ye shall say unto this mountain, Remove hence to yonder place; and it shall remove; and nothing shall be impossible unto you." (Matthew 17:20)*

Christ used the mustard seed to reveal to the disciples their unbelief. They did not have even so much faith as the tiny mustard seed. If they had had even this small amount of faith, they could have done great works like removing mountains.

But Christ had more to say. He went on to give a caveat in the next verse, *"Howbeit this kind goeth not out but by prayer and fasting." (Matthew 17:21)* Faith will move mountains, but when dealing with devils, something more than simple faith is needed. For this, prayer and fasting must be added to faith.

This is seen also in Christ's temptation by the devil in the wilderness. Matthew 4 says, *"Then was Jesus led up of the Spirit into the wilderness to be tempted of the devil. And when he had fasted forty days and forty nights, he was afterward an hungred. And when the tempter came to him, he said, If thou be the Son of God, command that these stones be made bread." (Matthew 4:1-3)*

As both God and man, Christ gave us a wonderful example of overcoming the devil's temptation here. Had He been God only and not man, He could not have been tempted, for *"God cannot be tempted with evil." (James 1:13)* But because He was also man, Christ *"was in all points tempted like as we are, yet without sin." (Hebrews 4:15)* Because He was man like we are, Christ here demonstrated the perfect way to meet and defeat the temptation of the devil.

First, Christ fasted. Second, though the text does not say so explicitly, we can infer from His habits that He also prayed during these forty days. Fasting and prayer are two sides of the same coin. And third, we also know that Christ, as the perfect Son of God, had perfect and full faith. He thus manifested all three elements necessary for dealing with devils, faith, prayer, and fasting. Just as He said to the disciples, He demonstrated for us here, that it takes all three of these for this difficult task of dealing with devils.

Luke 17 also mentions the mustard seed as an illustration of faith. This passage is not talking about dealing with devils, but with people. The context here is forgiving people who offend us. And so fasting and prayer are not mentioned, but still faith is emphasized. Christ says of faith in this context, *"If ye had faith as a grain of mustard seed, ye might say unto this sycamine tree, Be thou plucked up by the root, and be thou planted in the sea; and it should obey you." (Luke 17:6)*

It takes faith to deal with people. It takes faith to forgive them when they have offended us. And it takes faith to help them overcome these sins.

We can be thankful that faith is something all Christians have to some degree. Romans speaks of the varying degrees of faith, *"Having then gifts differing according to the grace that is given to us, whether prophecy, let us prophesy according to the proportion of faith." (Romans 12:6)* Each Christian has a different proportion of faith, some great and some little. But everyone has some faith, and even small faith can do great things.

Even faith as small as a grain of mustard seed can move mountains and trees. This doesn't take great faith, but only faith as a grain of mustard seed.

In light of this, how much smaller than a grain of mustard seed must our faith be? We don't do great things, and so our faith must be even smaller than this. Christ rebukes us in Luke 12 for this small faith, saying, *"If then God so clothe the grass, which is to day in the field, and to morrow is cast into the oven; how much more will he clothe you, O ye of little faith?" (Luke 12:28)*

Indeed we are deserving of this rebuke. We are *"of little faith."* Our faith is very small, smaller even than a grain of mustard seed. We do not do the great things that we could and should.

Let us take this rebuke and profit from it. Let us grow in faith. At the very least, let us grow in faith to the size of a grain of mustard seed. This is all God requires in order for us to accomplish great things for His glory.

31
The Myrtle Tree

The myrtle tree is really a shrub and not a tree per se. It has no trunk, but rather several branches coming from its roots. However, it does have woody stems and even its lack of a trunk is a characteristic it shares with some true trees, and so we include it in this study of the trees of the Bible.

We will begin with the myrtle as it is mentioned in Isaiah 41:19. We looked at this verse in chapter 7 focusing on the box tree and in chapter 18 focusing on the fir tree. But here we will look at it with a focus on the myrtle tree. God says in this verse, *"I will plant in the wilderness the cedar, the shittah tree, and the myrtle, and the oil tree; I will set in the desert the fir tree, and the pine, and the box tree together." (Isaiah 41:19)*

The myrtle here is planted in the wilderness. This is not the natural place for a myrtle. The wilderness is a dry, barren place, incapable of sustaining plant growth. Yet the Lord plants and sustains the myrtle here, just as He also plants and sustains the fir, pine, and box in the desert. This teaches us of the sustaining power of God. When He plants a tree, or when He plants us, even if it is not in our natural habitat, He can sustain us and make us grow.

The myrtle is unique among the trees named in this verse. It is the only shrub, whereas the rest are all tall trees. It is also the only one that produces flowers. This should be a reminder to us that each of us is unique and special in God's sight. Though we may not have the same gifts and talents that others have, we do each have our own special gifts and talents that are just as valuable to God as those of others. We are special to Him in our own way. Like the myrtle, we may not be tall trees like those about us, but we may have flowers that they do not have.

This verse prophesies of the day when the Lord will reverse the curse that exists upon the land of Israel. This land used to be a fruitful and productive land, like one vast garden. But the Lord brought a curse upon it because of Israel's sin in turning away from Him. Today, as a result, this land is a dry desert. But one day, Christ will turn it back into a garden. He will plant the myrtle tree in this barren wilderness, and restore the land to fruitfulness and abundance.

The God Who can do this with physical land can do the same in the spiritual realm. He can plant a spiritual garden in the place of a spiritual wilderness.

One such spiritual garden is the nation of Israel. God has made it a spiritual garden for the wilderness of the world. He promised Abraham, *"And in thy seed shall all the nations of the earth be blessed." (Genesis 22:18)* This blessing, brought to all the nations of the earth through Israel, is Jesus Christ. Finding Him, the wildernesses of the nations find in Israel the respite and refreshment of a fruitful garden.

Another spiritual garden is the church. The Lord has made the church a garden for all the unsaved of the world. Many make the foolish choice of staying in the wilderness of the sinful world, but God has given them a garden in the church if they will but come to it. They will find in it a respite from their dry and barren lives.

The Lord has also given us, as individual Christians, a spiritual garden from the wildernesses of our lives. We are not immune to troubles. We still encounter wildernesses at times. But Christ has provided a garden in the Bible, a place where we can find refreshment when we are in a dry land.

Another place where the myrtle shows the removal of God's curse is in Isaiah 55. Here Isaiah prophesies of a future day, *"Instead of the thorn shall come up the fir tree, and instead of the brier shall come up the myrtle tree: and it shall be to the LORD for a name, for an everlasting sign that shall not be cut off." (Isaiah 55:13)*

This verse foretells, not just the removal of the curse upon Israel, but the final removal of God's curse upon the whole earth. After the fall, God cursed the entire earth, saying, *"Thorns also and thistles shall it bring forth to thee; and thou shalt eat the herb of the field." (Genesis 3:18)* Since then, thorns and thistles and briers have been a constant reminder of sin and of God's curse. But in the day of Isaiah's prophecy, God will remove this curse at last, and He will cause the beautiful myrtle to come up instead.

Again, the God Who does this in the physical realm also does this in the spiritual realm. He takes the enemies and troubles in our lives, the thorns and briers, and changes them into pleasant trees like the myrtle and the fir. Sometimes we see this change and are allowed to enjoy it in our lifetime. But for some enemies and troubles, we may not see this change in our lifetime. It may not come until the resurrection. But we can rest assured that it will come, just as surely as the physical removal of the curse will come.

We are given final assurance of this in the last phrase of this verse, *"It shall be to the LORD for a name, for an everlasting sign that shall not be cut off." (Isaiah 55:13)* It is an everlasting change. The world may be getting worse and worse now, getting overrun with thorns and briers, but one day the Lord will remove these and bring instead the myrtle and the fir. This change will be permanent, and *"shall not be cut off."*

Another mention of the myrtle tree is in the book of Nehemiah at the celebration of the feast of tabernacles. God's law commanded this feast, saying, *"Go forth unto the mount, and fetch olive branches, and pine branches, and myrtle branches, and palm branches, and branches of thick trees, to make booths, as it is written. So the people went forth, and brought them, and made themselves booths, every one upon the roof of his house, and in their courts, and in the courts of the house of God, and in the street of the water gate, and in the street of the gate of Ephraim." (Nehemiah 8:15-16)*

The people were to use these myrtle branches, and the other branches, to make booths, or tabernacles. These tabernacles were to commemorate the tabernacles in which they had dwelt during their wilderness journey and the great tabernacle in which God had dwelt. The booths were to be on their houses, in the courts of their houses, and in the courts of the house of God.

Today we are not to build literal tabernacles in our houses or churches. But we are supposed to make our hearts into tabernacles in which God can dwell.

Peter speaks of us being tabernacles when he refers to himself with these words, *"Yea, I think it meet, as long as I am in this tabernacle, to stir you up by putting you in remembrance; Knowing that shortly I must put off this my tabernacle, even as our Lord Jesus Christ hath shewed me." (II Peter 1:13-14)* II Corinthians also calls us tabernacles. *"For we know that if our earthly house of this tabernacle were dissolved, we have a building of God, an house not made with hands, eternal in the heavens." (II Corinthians 5:1)* While we are on earth, our hearts are living tabernacles because we, like the tabernacle of the Old Testament, are the dwelling place of God Himself.

Perhaps the most beautiful and significant occurrence of the myrtle tree is found in the book of Zechariah. In the opening chapter of this book, Zechariah recorded this vision. *"I saw by night, and behold a man riding upon a red horse, and he stood among the myrtle trees that were in the bottom; and behind him were there red horses, speckled, and white." (Zechariah 1:8)*

Who is this man among the myrtle trees, and who are the others behind him? Zechariah wondered the same question, *"Then said I, O my lord, what are these? And the angel that talked with me said unto me, I will shew thee what these be." (Zechariah 1:9)*

However the angel ceased to speak here, and the man among the myrtle trees, himself, answered. *"And the man that stood among the myrtle trees answered and said, These are they whom the LORD hath sent to walk to and fro through the earth." (Zechariah 1:10)*

He answered concerning those behind him, but he did not reveal his own identity. However in the next verse, his identity is revealed in the title by which he is called, *"And they answered the angel of the LORD that stood among the myrtle trees, and said, We have walked to and fro through the earth, and, behold, all the earth sitteth still, and is at rest." (Zechariah 1:11)*

He is none other than the angel of the LORD. Thus He is no mere man. This angel of the LORD is the same being that appeared to Moses in the burning bush and to Gideon at the threshing floor. He is none other than the preincarnate Christ Himself.

Christ is pictured here as riding a red horse. We are not told if the horse is naturally red, like a sorrel horse, or if it is stained red with blood. Either way, it is a picture of the blood of sacrifice. The animal sacrifices under the law of the Old Testament were very numerous. They fully occupied the priests, who would have been very bloody after a day's work of sacrifice.

Christ, riding a red horse, pictures this sacrificial blood. Only in this case, He is not only the High Priest, but He is also the sacrifice. He is the final and perfect sacrifice for our sins. It is His blood that pays for our sin once and for all. After His sacrifice, there is no longer any need for priests to sacrifice animals.

For this reason, when Christ rides a horse in the New Testament, in Revelation, it is a white horse. *"And I saw heaven opened, and behold a white horse; and he that sat upon him was called Faithful and True."* *(Revelation 19:11)* Under New Testament grace, the perfect sacrifice of Christ is complete, and there are no more animal sacrifices. The red horse is no longer pictured, but the white. This white horse is a symbol of victory. It shows that Christ has gained the victory on the other side of His sacrifice, victory over death, hell, and the grave.

Of the horses behind Christ, some are red, some are speckled, and some are white. They represent past followers of Christ, the Old Testament prophets, priests, and patriarchs.

Yet the symbolism doesn't stop here. Coupled with the Hebrew word for myrtle, this vision of Christ among the myrtle trees teaches us another truly beautiful truth.

First we must look at the Hebrew word for myrtle, *hadas* (הדס). This word is special in the Bible because its feminine form is *Hadassah*, the Hebrew name of Esther. It is Esther's book, coupled with Zechariah's vision, that teaches us our important truth.

The book of Esther never mentions the name of God. Yet this does not mean God is not there. In fact, He can be seen all throughout the book of Esther.

In the same way, before his vision, Zechariah could not see Christ. But this did not mean that Christ was not there. Once Zechariah's eyes were opened supernaturally with a vision, he saw clearly that Christ was there, among the myrtle trees.

178

This is similar to another occasion when the Lord opened a man's eyes. Elisha's servant feared the Syrian army because he could not see God's supernatural armies though they were all about him. Elisha prayed for him, *"LORD, I pray thee, open his eyes, that he may see." (II Kings 6:17)* Only then could the servant see what was there all along. *"And the LORD opened the eyes of the young man; and he saw: and, behold, the mountain was full of horses and chariots of fire round about Elisha." (II Kings 6:17)*

The same is true for us today. We cannot see Christ physically with us or even in us, as in the tabernacle we mentioned earlier. When we look in the mirror or when we get an x-ray made, we don't see Christ in physical form. But this in no way means that He is not here. He is here with us in just as real a way as He was with Esther and with Zechariah among the myrtle trees.

In a certain sense, Christ is still among the myrtle trees. The myrtle is a small, flowering, sweet-smelling shrub, which is also a very fitting description of the church. It is small, flowering, and sweet-smelling to Christ. Just as He was among the physical myrtle trees in the book of Zechariah, so He is among the spiritual myrtle trees, the churches, today.

Let us take comfort in this and be encouraged by it. Just as Christ was there among the myrtle trees, so He is with us, among our myrtle trees, today.

32
The Oak Tree

The Biblical oak tree is similar to, but not the same as, our North American oak trees. There are many varieties of oak, all in the broad Latin classification *Quercus*.

To discover the characteristics of the Biblical oak, we must now look at the nature of oak trees and the Hebrew terms for the Biblical oak. This study is very detailed, and so we will take it in very small bites.

First, for the nature of oak trees, we must understand that there are four primary taxonomical families of oak. There is the red oak, the white oak, the golden oak, and the ring-cupped oak. In the United States, we are familiar with two of these, the red oak and the white oak. These are similar to one another, but do have certain characteristics by which they can be distinguished.

Next, we must learn the definition of an oak tree, for it will be important to keep in mind throughout this study. It is a three-part definition. First, the oak tree bears acorns. Second, it has monoecious, or single, leaves and flowers. And third, its male catkins are pendant, or in other words, its spikes of flowers hang down. We must keep these three characteristics of the oak in mind as we begin this study, and we must use them not only to understand the oak, but also to evaluate other suggested trees.

The main tree suggested is the terebinth tree. Many commentaries say the Biblical oak tree is really the terebinth tree. I don't believe this to be the case because the terebinth is vastly different from the oak. Commentaries also say terebinth for other different trees as well. It seems that the terebinth is their default tree whenever they don't know what a tree is.

Using the definition of oak, we can examine the terebinth tree. First, it doesn't bear acorns. It is instead in the *Pistachia* genus and bears red berries. Second, it has pinnate leaves, or several leaflets on one stem, whereas the oak has single leaves. For these reasons, I believe the Biblical oak is not a terebinth tree, but a true oak tree.

Here we also need to establish some interesting facts about the oak that we will need to apply later in this study. One fact is that the oak tree is struck by lightning more than any other tree. Another is that for every ten thousand acorns produced, only one oak tree grows. In light of this, one tree may not even produce another tree.

Another interesting fact about the oak is that its wood is the only wood used for holding liquid, particularly white oak wood. White oak wood is full of tiloses, a gum-like substance. This waterproof wood has been used throughout history for making barrel staves, the thin slats of wood that make up a wooden barrel, as can be seen in an old ice cream maker. In fact, the white oak is so well-suited for this task and used so commonly for it that it has been called the stave oak.

A final interesting fact about oak trees is that the different kinds of oaks interbreed with one another very easily. For this reason, some kinds of oaks are very hard to distinguish from others. They have interbred and look very similar.

Having looked at the nature of the oak tree, we are now ready to turn to the Hebrew words that are translated "oak." At first glance, there appears to be four different words, *elon* (אלון), *allon* (אלון), *allah* (אלה), and *elah* (אלה). But some of these can be quickly and easily eliminated.

If we look at the actual Hebrew letters of the first two words, *elon* (אלון) and *allon* (אלון), we see that they are simply two forms of the same word. The same is true of the next two, *allah* (אלה) and *elah* (אלה). This helps us narrow our focus to just two Hebrew words, אלון (*elon* or *allon*) and אלה (*allah* or *elah*).

The first form of the first word is easy to eliminate as the oak. It is always translated "plain" or "plains" in the King James Version. For example, Genesis 12 uses this word describing Abram's journey. *"And Abram passed through the land unto the place of Sichem, unto the plain of Moreh." (Genesis 12:6)* Some commentaries say that this should be translated as the "oak of Moreh." But this does not always fit the context of the verse. If we look at its occurrences throughout the Bible, it is always translated "plain," and "plain" always matches the context. We can safely conclude that this word is properly translated "plain," not "oak."

The word "plain" is also sometimes used to translate the word *biqah* (בקעה), which can mean either valley or plain. So not every occurrence of the word "plain" is from the word *elon* (עלון), but every occurrence of the word *elon* is translated "plain." So this Hebrew word *elon* clearly refers to a plain, and not to the oak tree, and thus it does not fall within our study of the oak tree.

This brings us to just three remaining words, *allon* (אלון) and the two forms of אלה, *allah* and *elah*. Just because three different words are used, this doesn't necessarily mean that there are three different kinds of oaks. We even have this in English. As we mentioned earlier, the white oak is the same tree as the stave oak, simply viewed from a different point of view. So I believe the different names of the oak are used at different times according to the speaker's point of view.

We will now examine each of these remaining three names. We will first look at the least common, *allah* (אלה), which is used only once in the Bible. This is the same form as the name of the Muslim deity, Allah. However, Muslims claim that their word Allah is different from the *allah* that means oak. They don't want their god to be merely an oak tree, and so they say the word is different.

They say that the name Allah means "the only one to be worshiped." Of course, we would agree that God is the only one to be worshiped, but they are not talking about our God, Jesus Christ. Instead they have a very different god. They further claim that this word is pronounced differently than the word for oak. They say that the name of their god should really be pronounced as ēla, not äla. However, they contradict themselves when they come to the word *elah*. They pronounce this word just the same as we do.

These Arabs also contradict themselves when they trace the origin of their language. They claim that they got their language from the Aramaic language. But in Aramaic, as in Hebrew, *allah* means "oak." It is also pronounced the same. In other words, when you come down to it, the name *Allah* really does mean "oak." *Allah* is not a true God, but merely a false and manmade deity drawn from the word for the oak tree.

Muslims also are faced with a problem regarding the origin of the name of their god. This name was not used until A. D. 400. Before that, the word *Allah* did not exist as a name in their language. This is a great problem for the validity of the Muslim god because all of the Bible was written before A. D. 400. In a Sanskrit writing, the name did appear once as the name of a goddess, but this pagan origin is even worse than no origin.

Allah must either be a name they have made up, or else it must come from pagan origins. Either way, we can conclude that *Allah* is a manmade god. When the Muslims changed the word *allah* into a proper name for their god, they may have changed the meaning to be "the only one to be worshiped." But if they did, this is a meaning that they have attached to it, not a meaning that is inherent in the word *allah* itself. The word *allah* means simply "oak." Muslims may try to trick people into thinking the word for their god and the word for oak are different words, but they are really the very same word. *Allah* is no greater than a mere oak tree, the true "only one to be worshiped" is the God of Heaven.

We will here look at the only place in the Bible where *allah* is mentioned. Joshua 24 records the covenant of the children of Israel to obey the law of God. *"And the people said unto Joshua, The LORD our God will we serve, and his voice will we obey. So Joshua made a covenant with the people that day, and set them a statute and an ordinance in Shechem." (Joshua 24:24-25)* To commemorate this covenant, *"Joshua wrote these words in the book of the law of God, and took a great stone, and set it up there under an oak, that was by the sanctuary of the LORD." (Joshua 24:26)*

Under this oak in Shechem, Joshua placed a stone to testify of the people's covenant to keep the law of God. The law of God was something very highly prized and very sacred. It was not to be changed or tampered with, but strictly followed and obeyed. The stone under the oak tree was a constant reminder of this unchangeableness of God's law. It was a great stone, a very large stone that was hard to move. To change the law of God, the people would have had to remove this very large stone. The message was clear. The law of God must not and could not be changed.

Another reference that possibly ties into this one is found in Genesis 35, a few hundred years earlier than Joshua. Here Jacob had been with Laban, his father-in-law, and now God commanded him, *"Arise, go up to Bethel, and dwell there: and make there an altar unto God, that appeared unto thee when thou fleddest from the face of Esau thy brother." (Genesis 35:1)* In obedience, Jacob gathered his new family and began his journey home to where his father Isaac and brother Esau were.

But before Jacob and his family reached Bethel, they came to the town of Shechem. And here they did an interesting thing. *"And they gave unto Jacob all the strange gods which were in their hand, and all their earrings which were in their ears; and Jacob hid them under 'the' oak which was by Shechem." (Genesis 35:4)* [emphasis added]

Just as in the account of Joshua, we read here of an oak in Shechem. The Bible doesn't say whether or not they are the same oak. This one here is called **the** oak. Joshua's is called **an** oak, which might suggest an entirely different oak. Indeed, there may have been many oaks in Shechem. But it is at least a possibility that this oak could be the very oak under which Joshua would set up the stone years later.

The Hebrew word behind this oak here in Genesis is *elah*, but the word in Joshua for possibly the same tree is *allah*. This is where we must consider the speaker's point of view. Joshua's point of view had to do with the law of God. Under this tree, he erected a stone as a memorial of the covenant to keep the law of God. This law of God was something proper and right, something that Christ supported, and something that was Christ-honoring.

This is the only use of the oak in the Bible that is connected with the law of God in any way, and this is the only place in the Bible where the oak is called *allah*, which carries the meaning "the only one to be worshiped." Indeed, not the Muslim Allah, but the God of Heaven is the only one to be worshiped. It was this true God and His Word that Joshua was honoring under the oak tree, and accordingly the name *allah* was used.

However, Jacob's point of view was different. Under possibly the same tree, he buried idols, something impure, something that Christ did not approve of, and something that did not honor Him. The word *elah* used for the oak in this account reveals this opposite point of view.

The term *elah* comes from the Hebrew root *el*, which means "strength." This word *el* was used by the Hebrews as their term for "god" because a god is supposed to be strong and powerful. Just like our English word, the same word could be either the true God or a false god.

We see this word *el* in this same account of Jacob just a few verses later. *"And he built there an altar, and called the place Elbethel: because there God appeared unto him, when he fled from the face of his brother." (Genesis 35:7)* Bethel by itself has the *el* ending and means "house of God." But here another *El* is added so that the word for God is used twice. This makes Elbethel, which means "the God of the house of God." In this context, *El* refers to the one true God.

But the Hebrew word *el* could refer to any god, even a false god. And so it reveals Jacob's point of view, that of burying false gods, as opposed to Joshua's point of view, that of honoring the true God.

Having covered the word *allah* (אלה), we will now move on to the two main words for oak, *allon* (אלון) and *elah* (אלה). The first of these words is masculine, *allon*, while the second is feminine, *elah*. This distinction I believe to be the reason that they are both used for the oak in the Bible, as it appears in different contexts and from different points of view. Though these Hebrew words are not really adjectives but nouns, for ease of communicating, I will use the terms "*allon* oak" and "*elah* oak."

We will begin with the *allon* oak, the one with the masculine word. It is used when the Bible is speaking of the masculine qualities of the oak, its strength and power. It is the used to describe the oak as used for manly things, like oars.

The first occurrence of this word is found in the very same passage as Jacob's burial of the idols under the oak in Shechem. To get the context, we will read the intervening narrative. *"And they journeyed: and the terror of God was upon the cities that were round about them, and they did not pursue after the sons of Jacob. So Jacob came to Luz, which is in the land of Canaan, that is, Bethel, he and all the people that were with him. And he built there an altar, and called the place Elbethel: because there God appeared unto him, when he fled from the face of his brother."* (Genesis 35:5-7)

In the next verse, we come to the *allon* oak. *"But Deborah Rebekah's nurse died, and she was buried beneath Bethel under an oak: and the name of it was called Allonbachuth."* (Genesis 35:8) In the Hebrew name given to this oak, *Allonbachuth*, we can identify our masculine name for the oak, *allon*. Putting this with the word *bachuth*, which means "weeping," the entire name, *Allonbachuth*, means "oak of weeping."

This oak of weeping is a different oak than the one under which Jacob buried the idols, for it is in Bethel, not Shechem. Jacob and his family had by now left Shechem and traveled on to Bethel. So we cannot connect Deborah's burial with the other burials in Shechem. She died in Bethel along the journey, and this oak was used simply as a place of burial for her.

The *allon* oak is also mentioned in Hosea. Hosea here speaks the word of the Lord, *"My people ask counsel at their stocks, and their staff declareth unto them: for the spirit of whoredoms hath caused them to err, and they have gone a whoring from under their God. They sacrifice upon the tops of the mountains, and burn incense upon the hills, under oaks and poplars and elms, because the shadow thereof is good: therefore your daughters shall commit whoredom, and your spouses shall commit adultery."* (Hosea 4:12-13)

The trees in this passage are all great and tall trees. As such, they have a large shadow and provide great shade. This is a place where the whores and idol worshipers of these verses can get out from the heat of the sun.

But it is also a dark place. The ones who come here are whores and idol worshipers. They come here to practice evil. And the Bible says of such people, *"Men loved darkness rather than light, because their deeds were evil."* (John 3:19) This is the case with these men. They come to this dark and shady place thinking that here they can hide their sin.

It is interesting that these ancient sinful Israelites are not the only people to worship under the oak tree. The oak has been worshiped as an idol for centuries. The Druids worshiped it, as did many pagan peoples. It is even said that ancient heathens consecrated their false god Jupiter under an oak tree.

It is also interesting to note the location of these trees and this idol worship. It was done *"upon the tops of the mountains"* and *"upon the hills."* Again, throughout history, a lot of worship has been done on hilltops. American Indians even built large mounds on which to worship. It seems that somehow, they thought this would bring them closer to the objects of their worship, such as the sun.

Still today, churches are often built on hills, especially weak churches that want to be noticed and want to be prominent. True churches with the true Gospel of Christ do not need this elevation. They do not need their building to be prominent, because their people are prominent and well known as Christians. This is the right kind of prominence, which Christ desires, the prominence of His people, not of a building. But weak and shallow churches want men to see how big and how nice they are. They want the place of prominence high on a hilltop.

Zechariah also mentions the *allon* oak tree. *"Howl, fir tree; for the cedar is fallen; because the mighty are spoiled: howl, O ye oaks of Bashan; for the forest of the vintage is come down."* (Zechariah 11:2)

Here he refers to the *"oaks of Bashan."* The word "Bashan" literally means "soft, fertile soil." So the phrase *"oaks of Bashan"* could be read "oaks of soft, fertile soil." Oaks growing in such rich soil would grow fast and strong. We all know this from experience. Trees, or any crops such as corn, grow best in soft and rich soil. They don't grow very well in hard soil. These oaks of Bashan are grown in soft, fertile soil, and so they are goodly trees, flourishing and strong.

The phrase *"oaks of Bashan"* appears three times in the Bible. The oak tree was a famous resource of the land of Bashan.

Indeed Bashan was known for its oaks as much as Lebanon was known for its cedars.

Yet, in these verses of Zechariah, these famous and mighty trees are told to howl because the cedars, mighty trees like themselves, are fallen. This is used by the Lord to show Israel how much they have fallen. They were once strong in God's grace, but now they have fallen.

Indeed the downward progression of Israel can be clearly seen in three places in Scripture. First, early in their history, about 1400 B.C., when they had just come out of the wilderness, Joshua put before them this choice, *"And if it seem evil unto you to serve the LORD, choose you this day whom ye will serve; whether the gods which your fathers served that were on the other side of the flood, or the gods of the Amorites, in whose land ye dwell: but as for me and my house, we will serve the LORD."* *(Joshua 24:15)*

The people gave a faithful and loyal response. *"And the people answered and said, God forbid that we should forsake the LORD, to serve other gods."* *(Joshua 24:16)* With this spirit, the nation was strong and mighty.

But if we move ahead five hundred years, we come to a second and very similar choice. By now, Joshua was off the scene, and Elijah presented the choice to the people. *"And Elijah came unto all the people, and said, How long halt ye between two opinions? if the LORD be God, follow him: but if Baal, then follow him."* *(I Kings 18:21)*

This was almost the very same question Joshua had set before the people. But this time, their response was much different. *"And the people answered him not a word."* *(I Kings 18:21)* The first time, they had said, *"God forbid."* But now they had not one word to give Him. They didn't say anything.

The third time this choice was presented to Israel is recorded in John 19. Pilate was the one this time who set the choice before the people. *"Pilate saith unto them, Shall I crucify your King?"* *(John 19:15)*

This time, the people took the final step downward. *"The chief priests answered, We have no king but Caesar."* *(John*

19:15) They disowned Christ as their King and proclaimed someone else, Caesar, as their king.

How far they had fallen! They went from saying, *"God forbid that we should forsake the LORD,"* to keeping silent, and finally to hailing another as their king. This falling is what Zechariah is talking about, and what he tells the oaks of Bashan to howl about. The people of Israel once were very high and very strong, but now they have fallen far from where Christ wants them to be.

This should be a warning to us as Christians. Spiritual falling happens slowly, almost imperceptibly. For this reason, we must be very alert and on guard that it does not happen to us.

Another passage that speaks of the *allon* oak is Amos 2. *"Yet destroyed I the Amorite before them, whose height was like the height of the cedars, and he was strong as the oaks; yet I destroyed his fruit from above, and his roots from beneath."* *(Amos 2:9)*

Both the oak and the cedar are used here to describe the Amorite nation. These are both very mighty trees. As we see in this verse, the cedar is tall, and the oak is strong. Together, they give a complete picture of the Amorites. They were both tall and strong. Here the masculine connotation of the *allon* oak is seen. It speaks of might, power, and strength.

Yet for all the Amorites' height and strength, God still destroyed them. He said, *"Yet I destroyed his fruit from above, and his roots from beneath."* He destroyed both the branch and the root. He didn't stop with just the branch above, but He went all the way to the root. He destroyed the Amorites both top and bottom.

The Amorites were some of the Canaanite people whom the ten unbelieving spies feared when they first entered the land of Canaan. These spies reported, *"The land, through which we have gone to search it, is a land that eateth up the inhabitants thereof; and all the people that we saw in it are men of a great stature."* *(Numbers 13:32)*

Yet in spite of their great stature, God was fully able to destroy them, and indeed He did destroy them. He destroyed them so completely that their name and tribe disappeared from the face of the earth, seemingly around the time of Solomon. In fact, if Amos had not been retelling the history of the past, the name "Amorite" would not have even come up in his book, for they were long destroyed by Amos' time.

This complete and total destruction of the Amorites, tall and strong though they were like the cedars and oaks, shows the even mightier power of God. Compared with His almighty power, even the mighty, masculine *allon* oaks, and the Amorites they represent, are weak. It doesn't matter how tall or strong anyone is. The Lord can destroy even the most mighty, both branch and root, both top and bottom.

Isaiah also mentions the *allon* oak in chapter 44 as being used to make idols. *"He heweth him down cedars, and taketh the cypress and the oak, which he strengtheneth for himself among the trees of the forest: he planteth an ash, and the rain doth nourish it."* *(Isaiah 44:14)* We have looked at this passage in earlier chapters, for it mentions several trees that we have already covered. But here we will focus on the oak tree.

It has the most beautiful and durable wood of those in this verse. But tragically, as we see here and also throughout history, the oak tree is connected with pagan worship. In fact, it is so connected with it that one of the best known pagan peoples, the Druids, were named for their worship of the oak tree. The name "Druid" has an interesting etymology. One word it comes from is the old Irish word *drul*, which means "sorcerer." A related word is the Welsh *dryw*, which means "seer." Still another word is the Greek *drus*, which means oak tree. All these words combine to form the ancient word "Drywids," or its modern form "Druids," meaning "oak-seers." Traces of these ancient roots can still be found today. For example, the modern Irish word for oak is *doir*. Many Irish words come from this root, including Derry and Doire. Kildare also comes from this root and means "church of the oaks."

Obviously, based upon their own etymology, the Druids worshiped the oak tree. Historians also say that they worshiped not only the tree itself, but even the mistletoe that grew in it.

In keeping with this pagan reputation, the oak tree is often used as a picture of evil in the Bible. This is the opposite of most of the trees we have studied. Trees like the cedar, the fig, and the olive are almost always pictures of good. But the oak, while it can occasionally picture good in some places, usually pictures evil.

In Isaiah 2, the *allon* oak is mentioned again, and again the phrase *"oaks of Bashan"* is used. Just as in Zechariah 11, these oaks of Bashan are a picture of strength, used to demonstrate God's even greater strength. *"The lofty looks of man shall be humbled, and the haughtiness of men shall be bowed down, and the LORD alone shall be exalted in that day. For the day of the LORD of hosts shall be upon every one that is proud and lofty, and upon every one that is lifted up; and he shall be brought low: And upon all the cedars of Lebanon, that are high and lifted up, and upon all the oaks of Bashan." (Isaiah 2:11-13)*

The cedars here are high and lifted up, and the oaks are strong. They are proud in their own strength. This is another expression of the masculinity of the *allon* oaks, their pride and strength. But they will brought down for this very pride of lifting themselves up, for the Lord will bring down all those who lift themselves up.

This passage is speaking of the last days that are coming. Like these oaks, the kings and rulers and presidents of the earth may think that they're strong now. But to the Lord's eyes, they are weak, and He will easily bring them down. He will root them out like He did the Amorites. Together the proud oak and lofty cedar will fall. They will be destroyed both root and branch, so that they are no more in the land.

The *allon* oak is also found in Ezekiel 27. We looked at the fir and cedar trees in this passage in previous chapters. *"They have made all thy ship boards of fir trees of Senir: they have taken cedars from Lebanon to make masts for thee. Of the oaks of Bashan have they made thine oars; the company of the Ashurites have made thy benches of ivory, brought out of the isles of Chittim." (Ezekiel 27:5-6)*

The Tyrians, or Phoenicians, are the people spoken of here. They were famous for shipbuilding, as they either were the first shipbuilders or made early improvements in shipbuilding that greatly advanced the industry. In this verse, we find that they made the ship masts of cedar and the ship boards of fir, but they made the oars of oak, of the oaks of Bashan. It is here that we now focus our attention.

Oak is a very strong, durable wood. For shipbuilding, this durability and strength are desirable, but only for certain parts of the ship, as the Tyrians demonstrated with their choices of different woods. Ships need to be strong, but only strong enough to accomplish the purpose of sailing for which they are designed. Spruce and pine, and even fir as in this passage, are light woods that allow the ship to ride high in the water and therefore sail close to the shore. Were the entire ship to be made of heavy oak, it would be strong, but too heavy for sailing close to the shore. Thus its weight would prevent it from sailing to the coast and accomplishing its purpose.

So it is not for the ship boards, but for the oars, that oak is most desirable. Oars get a lot of wear, being continually rubbed against their sockets and constantly handled by men. The lighter woods, such as spruce, pine, and fir, would wear down very quickly under such use. So oak is the ideal wood for this task, allowing the ship to accomplish its task of sailing with minimal maintenance.

Because of the hardness of the oak, these oars were carved by the Tyrians while the oak wood was still green. Once oak wood dries, it becomes very hard, and shaving it is very difficult. But carving it while it was green made the task much easier for the Tyrians.

These Tyrians even had an ingenious method of storing oak wood and keeping it green through the winter. They wouldn't work much once it got cold in the winter, and so before it got cold, they would take their green oak wood and bury it in mud and water. This would keep it green so that when they dug it up in the spring, it was still green and soft and easily shaved.

To conclude our study of the *allon* oak, we will now look at some passages where the word "Allon" appears as a proper name. We can recognize these places by the capitalization of "Allon" in the King James Version. While there are no capital letters in Hebrew, the capitalized English word indicates that the word is used as a proper name. We have already seen one of these,

Allonbachuth, "the oak of weeping" where Deborah, Rachel's nurse, was buried.

In I Chronicles, Allon is listed as a name by itself. This time it is not the name of a place, but of a man. *"And Ziza the son of Shiphi, the son of Allon, the son of Jedaiah, the son of Shimri, the son of Shemaiah." (I Chronicles 4:37)* This chapter lists men of the tribes of Judah and Simeon. Allon is listed as a man in the tribe of Simeon. Perhaps this man was strong and mighty like an oak tree, or maybe this was just his name without any direct connection to the oak. We don't really know. For references like this, when the name of a tree is a proper name, we must always remember that the name may or may not be directly connected with the tree. However, unlike the word *elah* which we will come to in a moment, the word *Allon* used as a proper name is almost always directly connected with the oak tree.

Another reference in which Allon is used as a proper name is in Joshua. Here it is the name of one of the towns that formed the boundary of the inheritance of the tribe of Naphtali. *"And their coast was from Heleph, from Allon to Zaanannim, and Adami, Nekeb, and Jabneel, unto Lakum; and the outgoings thereof were at Jordan." (Joshua 19:33)* Again we must be careful in our interpretation. This could be a town abounding with oak trees, but we don't know that for sure.

This concludes our study of the *allon* oak. It is the masculine oak, and symbolizes strength, might, and power.

As we move to the *elah* oak, the feminine word, we must pause briefly to look at this Hebrew word. It has caused some confusion because it is the same word translated elm tree and also teil tree in the Bible. I believe this is because the word *elah* literally means "strength," and it speaks more about the strength of the tree than about its species. There can be strong elms, strong teils, and strong oaks, all of which are called by the term *elah*.

We do a similar thing in English. We have the white oak, the white elm, and the white pine. These are all described by the name "white," but that doesn't mean they are all the same tree. This is merely a descriptive term, describing a common characteristic that these trees share, whiteness. I believe this to be the case with the term *elah*. It describes a common characteristic, strength, that is shared by the elm, the oak, and the teil.

The same thing applies when we look at "Elah" as a proper name. I believe it refers to strength, not necessarily to the oak tree. Several men were named Elah in the Bible. There was Elah the son of Caleb, Elah the son of Uzzi, and Elah the son of Baasha, who was one of the kings of Israel. This name "Elah" does not necessarily mean that they were like oak trees. Instead it probably was chosen for its meaning of "strength" and means that they were all strong men.

A famous use of the proper name "Elah" is in the name "Valley of Elah" where David killed Goliath. *"And Saul and the men of Israel were gathered together, and pitched by the valley of Elah, and set the battle in array against the Philistines." (I Samuel 17:2)* Again, this is not necessarily a valley full of oaks or elms, but rather a valley of strength, might, and power. Indeed, it was the place where the power of Israel was pitched against the power of the Philistines, and deeper still, the power of God against the power of Satan. David's victory here dramatically displayed the strength of God as triumphant over the strength of the devil.

We will now look at some specific occurrences of the *elah* oak. The first occurrence is the one we have already looked at, where Jacob buried his family's earrings and idols under an oak tree in Shechem. The Bible doesn't say if these items were buried willingly or unwillingly.

Perhaps the burial was done by Jacob against the will of his family. This might indicate why the exact location of this oak was not given. It was said to be *"by Shechem," (Genesis 35:4)*, but no more specific than that. Perhaps Jacob buried the idols and earrings in an unknown place so that his family could not get back at them.

On the other hand, both the oak and the idols were connected with pagan religion and superstition. If Jacob's family truly believed in the superstition connected with these gods, they may have been afraid to venture near and defile this place by digging up their possessions. This superstition alone might have been enough to keep them from returning.

In any case, they didn't stay there in Shechem where the idols were, and where they might be tempted to dig them back up. The next verse says, *"And they journeyed."* (Genesis 35:5) They immediately moved on to Bethel, the house of God.

This is an important lesson for us to learn. If we "bury idols," then we must turn away from them and go toward Christ. We cannot linger where the idols are, lest we be tempted to turn back to them. Instead, we must go to Christ.

Another place where the *elah* oak is mentioned is in Isaiah 1, where God pronounces judgment upon the sinners of Zion. *"For they shall be ashamed of the oaks which ye have desired, and ye shall be confounded for the gardens that ye have chosen. For ye shall be as an oak whose leaf fadeth, and as a garden that hath no water. And the strong shall be as tow, and the maker of it as a spark, and they shall both burn together, and none shall quench them." (Isaiah 1:29-31)*

Again, we find the oak connected with idolatry. The sinners of this verse have desired and worshiped idols of oak. Perhaps they have been trusting in the might and power of the oak tree. But now, when God arises to judge them, these oak idols are of no help to them. God says, *"The strong shall be as tow, and the maker of it as a spark, and they shall both burn together, and none shall quench them." (Isaiah 1:31)*

The tow referred to in this verse is hemp. Our word for towing vehicles comes from this same word, for this task was originally done with hemp rope. We are all familiar with the flammability of this substance. It catches on fire very easily and burns up very quickly. One match, or even a little spark, causes it to flame up and the whole rope to be consumed very rapidly.

This is what God says will happen to the oaks of this passage. Though naturally oak wood is strong and durable and not easily burned, this mighty oak wood, because it has been used for idols, will be burned up quickly and easily like hemp rope. And not only will it be burned, but the makers of the idols will also be burned, as the verse says, *"and the maker of it as a spark."* Both the idols and their makers will be burned up together by the One Who holds the match, Jesus Christ.

The *elah* oak is found in a brighter context in Judges 6, where the angel of the LORD appeared to Gideon. This passage begins, *"And there came an angel of the LORD, and sat under an oak which was in Ophrah, that pertained unto Joash the Abiezrite: and his son Gideon threshed wheat by the winepress, to hide it from the Midianites." (Judges 6:11)* It is significant and not without cause that this angel came and met Gideon under an oak tree, as we shall see.

Gideon's name means "tree feller," one who cuts down trees. Names in the Bible were often given as descriptions of the persons to whom they were attached. In this case, it is very probable that tree felling was Gideon's main occupation. As a tree feller then, Gideon would have immediately recognized the significance of the angel being under the oak tree. He would have known that the oak was a symbol of power and strength. Seeing the mighty angel under the mighty oak tree, Gideon would have seen clearly the message that God was giving him, that He had almighty power to accomplish the task He was about to command Gideon to do.

This assurance of power was something Gideon needed. We know that he was threshing wheat here at the winepress *"to hide it from the Midianites." (Judges 6:11)* Apparently, he was by nature timid and afraid, and was not know for his courage and strength. When he threw down the altar of Baal and cut down its grove, he did it by night because he feared the people who worshiped Baal.

To assist this timid and fearful young man, the Lord sent a mighty angel and placed him under a mighty oak. With these two combined, even the timid Gideon, being a tree feller, knew that this was something to which he should listen. As he did, the angel told him that he would destroy the Midianites as one man.

Even with all this reassurance, Gideon still was a little hesitant. He listened to the mighty angel under the mighty oak, and then gathered his men. But he still had to make sure one more time that his course was right, that this was indeed what the angel wanted him to do. It was for this reassurance that he spread out his fleece and received the Lord's answer that it was the right course. Gideon, as a humble and weak man, received God's power under the oak.

However, as we shall see in the final occurrences of the *elah* oak, this was not the case with many others in the Bible. Those who were powerful in themselves and proud of their power were often found under oak trees, not to receive God's power, but to receive God's judgment. For the oak is also a picture of judgment. Those unwilling to admit their weakness must suffer this judgment, rather than be given God's power like Gideon.

It seems to be a pattern throughout the Bible that when a man is himself under an oak tree, he is receiving judgment. This is why it is important that Gideon was not under the oak tree. He was not claiming to be powerful, and he was not receiving judgment. It was the truly powerful angel who was under the oak tree in his account, and no judgment was needed. But when we see a mere man under an oak tree, a man who thinks himself powerful enough to identify with the oak, this is a sign of judgment.

The first of these is Absalom, the son of David. Absalom thought himself to be powerful and important and tried to take over the kingdom of Israel. To do so, he turned many of the people away from David his father and led them in a battle against David's men. His "judgment under an oak tree" is recorded in II Samuel. *"And Absalom met the servants of David. And Absalom rode upon a mule, and the mule went under the thick boughs of a great oak, and his head caught hold of the oak, and he was taken up between the heaven and the earth; and the mule that was under him went away."* (II Samuel 18:9)

Like in the account of Gideon, the Bible could have used the general term "tree" here, but instead it uses the specific word "oak." No word is ever used arbitrarily in the Bible. I believe the oak is used in passages such as these because of its symbolism of power and of judgment, so that we can see the contrast between such men as Absalom and Gideon.

Here in this passage, the proud Absalom was now hung between heaven and earth, between the mighty power of the oak and the lowly earth. And here, under the oak, his judgment came from the hand of Joab, David's captain. *"And a certain man saw it, and told Joab, and said, Behold, I saw Absalom hanged in an oak. . . . Then said Joab, I may not tarry thus with thee. And he took three darts in his hand, and thrust them through the heart of Absalom, while he was yet alive in the midst of the oak." (II Samuel 18:10,14)* Thus Absalom was judged in the oak tree, by Joab.

Was Joab correct in killing Absalom here? This is a question that has been debated for centuries. David had strictly commanded Joab and his other captains not to kill Absalom. It was for this reason that the first man to find Absalom in the oak did not kill him. He knew the king's commandment and would not go against it. Instead of killing Absalom, he went and told Joab where Absalom was.

On the other hand, Absalom was a murderer, and these two armies were at war. Perhaps Joab was acting to protect the people. Had he let Absalom live, he might have fully turned the kingdom away from David and endangered the safety of all of Israel. Perhaps, but even this is not fully satisfactory. God was perfectly capable of protecting Israel, and Joab perhaps should have left the safety of Israel in His hands.

Whether Joab was right or wrong in this case, I don't know. But we do know one thing, that our sins will find us out one way or another. We will be judged for them either in heaven or on earth. Absalom was judged for his sin here by the hand of Joab, but God didn't need Joab. The important thing is that Absalom was judged at all. We also know that Joab was later judged for his own sins.

Another "judgment under an oak tree" is recorded in I Kings 13. The one judged here was a disobedient man of God. He was not a false prophet, but a servant of the true God. Yet he was disobedient.

The Lord had sent this man on a mission and told him very clearly not to drink bread or eat water. But after he had fulfilled his task and was returning home, another prophet tried to bring this man back to eat bread with him. He *"went after the man of God, and found him sitting under an oak: and he said unto him, Art thou the man of God that camest from Judah? And he said, I*

am. Then he said unto him, Come home with me, and eat bread."
(I Kings 13:14-15)

He found this man of God under an oak tree. For some reason, the man had stopped. He had not fully completed the task God had for him, but stopped for a time. Had he continued on, he never would have faced this temptation. But just as an oak had stopped Absalom in his course, so an oak stopped this man in his course.

Tragically, in spite of the clear command of the Lord, he turned back to eat bread with this prophet. For his disobedience under the oak tree, and lifting his own desires above God's command, he was judged. The Bible records his gruesome judgment, *"And when he was gone, a lion met him by the way, and slew him: and his carcase was cast in the way, and the ass stood by it, the lion also stood by the carcase."* (I Kings 13:24)

God places a high importance upon obedience. It is one of the chief things He wants from us. God once made this clear to Saul, saying through Samuel, *"Behold, to obey is better than sacrifice."* (I Samuel 15:22) We can make sacrifices to God, but what He really wants is our obedience. This is far better than any sacrifice we could give.

This very point ties into the next "judgment under an oak tree," that of Saul. Saul wanted to offer a sacrifice to the Lord. But according to God's law, he had to wait for Samuel the priest, for only the priest could offer a sacrifice. However, Saul grew tired of waiting, and so he offered the sacrifice himself.

Yes, sacrifice is important, but obedience is more important. We should never disobey God in order to do sacrifice. In this case, Saul was not the one to do sacrifice. Samuel was. Saul should have waited no matter how long it took for Samuel to get there.

For his disobedience, not only here but on numerous other occasions, Saul was eventually judged under an oak tree. At his death, *"they arose, all the valiant men, and took away the body of Saul, and the bodies of his sons, and brought them to Jabesh, and buried their bones under the oak in Jabesh, and fasted seven days."* (I Chronicles 10:12) The reason for this judgment is given in the next verse. *"So Saul died for his transgression which he committed against the LORD, even against the word of the LORD, which he kept not, and also for asking counsel of one that had a familiar spirit, to enquire of it." (I Chronicles 10:13)*

Just as these men fasted and mourned for Saul, so we at funerals tend to remember the good things of a person's life, and their faults are usually buried with them. This ties back into the first oak we looked at, where Jacob buried the idols of his family. Sins and faults, once they are dealt with either through confession or judgment, are all buried under the oaks.

This concludes our look at the *elah* oak, the last Hebrew word translated "oak." This is a sobering place to end, at the judgments upon Saul, the disobedient man of God, and Absalom. But let this be a warning to us. Let us not be like these proud and disobedient men. Rather let us be like the humble Gideon, who instead of receiving judgment under the oak, received God's power upon his life. May this be the case with each one of us, that we might humble ourselves so that God's almighty power, symbolized by the oak tree, might fill and empower us for His service.

33
The Oil and Olive Trees

The term "oil tree" appears only once in the King James Version of the Bible. This is in Isaiah 41:19, a verse we have looked at for other trees. Here the Lord says, *"I will plant in the wilderness the cedar, the shittah tree, and the myrtle, and the oil tree; I will set in the desert the fir tree, and the pine, and the box tree together."* (Isaiah 41:19)

Conclusively defining this oil tree is a little difficult. But to begin, let's first look at the Hebrew. The Hebrew word translated "oil" is the word *shemen* (שֶׁמֶן), which literally means "fatness" and is often translated "oil." It is the common word used for lamp oil and anointing oil in the Bible. So the term *ets shemen* (שֶׁמֶן עֵץ) means "tree of oil wood," and is translated here "oil tree."

The pine has been suggested as the possible identity of the oil tree because it contains pine oil or pine tar, which burns easily and can be used as lamp oil. But the pine is already mentioned in the Bible, and even in this same verse. It would be improbable that the same tree would be called by two different names in the same verse. So I believe we can pretty easily eliminate the pine as the oil tree.

Another suggestion is the olive, whose oil is well known. The olive is also mentioned elsewhere in the Bible, and at first glace, it would seem we could eliminate it as an option too. But let us look a little deeper.

The New Testament also mentions the olive tree. There the Greek word for olive tree is *elaia* (ελαια). From this word comes the Latin name for the olive tree, *elaeagnus*. Our English name for the North American olive tree also comes from here. It is the name oleaster.

The oleaster is today more commonly known by its relatively newer name "Russian olive." This tree is an invasive species, and its leaves and wood are both very olive-like. Interestingly, the name "oleaster" was originally not attached even to this tree, but it was still attached to a type of olive tree.

In the New Testament, aside from the simple olive tree, another kind of olive is also mentioned. This is the wild olive tree. Its Greek word is *agrielaios* (αγριελαιος). This word still contains the root *elaia* (ελαια), or "olive," but it adds the prefix *agri* (αγρι), which means "field." Thus it literally says "field olive."

This tree that the New Testament calls a wild olive I believe to be the same tree that the Old Testament calls an oil tree. I wouldn't call this an absolute proven fact, but it is at least my hypothesis. The close connection between the oil and the olive in the Old Testament, coupled with the close connection between the wild olive and the olive in the New Testament, seems to point in that direction.

Let us now look at our verse in Isaiah and test this hypothesis. *"I will plant in the wilderness the cedar, the shittah tree, and the myrtle, and the oil tree; I will set in the desert the fir tree, and the pine, and the box tree together." (Isaiah 41:19)*

The oil tree here is listed last of the trees planted in the wilderness, just as the box is listed last of those planted in the desert. Under our interpretation of the oil tree as a wild olive tree, it is a small shrub and the smallest of all the trees in its list, just as the box is the shrub and the smallest in its list. This makes the order of the two lists consistent.

It may seem, however, that our interpretation of the oil tree gives us a contradiction. The word *shemen* (שמן) speaks of any cultivated oil and indicates that the tree has been cultivated for use. How can the oil tree be both a **wild** olive and a **cultivated** tree? But this only seems like a contradiction, and we find that it is not really one at all when we look at the context of the verse. This verse is describing, not man's planting of trees, but God's supernatural planting. This verse is intentionally contrary to nature because it is God, not man, Who is doing the planting. God, Who is all powerful, can cultivate even a wild olive tree.

We see the same thing with the shittah tree in the same verse. It is a very wild tree, with long, sharp thorns. Though its wood is useful, the tree itself is not. And so, no man would ever naturally want to plant and cultivate it. But God is not like man. He plants this wild shittah tree, and in the same way, He cultivates the wild olive tree.

This idea of cultivation I believe to be the key to understanding the use of the term *shemen* as related to trees. We will see this later as we come to more trees that use the word *shemen*.

An important subject while talking about the wild olive is that of grafting. The New Testament speaks of the wild olive as being grafted into the good olive. So in order for our hypothesis to stand, the oil tree must be able to be grafted into the olive tree.

Two dissimilar trees cannot be grafted into each other. If we look just at the Hebrew names for the oil and olive trees, they seem quite dissimilar, *shemen* (שֶׁמֶן) and *zayith* (זַיִת). But we must remember the similarity of their uses and their close connection in New Testament grafting.

Also, I have come across in my studies an ancient statement that the Phoenicians and Greeks grafted together an olive tree and an oil tree. This is a very interesting statement. It still doesn't say exactly what the oil tree was, but it does give some helpful clues. It conclusively rules out the pine, for a pine tree could never be grafted with an olive tree. It also rules out other suggestions, such as nut trees, which also couldn't be grafted with an olive tree.

In fact, this statement makes our hypothesis seem very likely. Only a tree that was very similar to the olive, such as a wild olive, could be grafted with it. While we can't say dogmatically that this is what the oil tree is, this is what all the evidence seems to indicate.

Another interesting note that also strengthens our position is the fact that, in Arabic the word for oil tree and the word for olive tree are the very same, *zackum*. Again, this points to a very close connection between the oil and the olive, making it highly probable that the oil tree is the wild olive tree.

Having covered the oil tree, we now turn our attention to the olive tree in the Bible. This tree is mentioned in about sixty passages in the Bible, and so we won't cover every mention of it. Some of these occurrences have very obvious meanings and don't need to be covered.

For example, I Chronicles says, *"And over the olive trees and the sycomore trees that were in the low plains was Baalhanan the Gederite: and over the cellars of oil was Joash."* *(I Chronicles 27:28)* This is a simple and clear statement of fact, that David set these men to be overseers of his olive yards.

We don't need to spend our time on clear verses such as these. Instead, we will focus on the passages that might be more obscure, and the ones that have a spiritual application which we can draw from them.

The Hebrew word for the olive is *zayith* (זִית), and the Greek word is *elaia* (ελαια). Besides *elaeagnus*, another Latin word for the olive tree that comes from this Greek word is *Olea*. Some of us no doubt remember from years ago a product called oleomargarine. This was a substitute for butter made from vegetable oil. Though it was not made of pure olive oil, its name was derived from the Latin word *Olea*, and ultimately from the Greek word *elaia* (ελαια).

We next need to establish some facts about the olive that will help us later on as we progress in our study. One fact is that every year, an olive tree needs to be pruned, and pruned severely. If it is not pruned, or if it is not pruned enough, it won't produce fruit.

Also like other fruits, green olives are not fully ripe. As they ripen, they turn black. This will be important when we look at the different kinds of oil they produce and the different uses of that oil.

Olives also cling very tightly to their tree. In order to remove them from the tree, it is necessary to beat the tree. They hold fast, and do not just fall off like other fruits.

If olives are to be made into oil, they must be processed into oil the same day they are removed from their tree. If they are left around for any length of time after their removal, they rot, and their oil becomes no longer useful.

The olive tree also likes long, hot, and dry days. This may seem strange to us. We usually think of trees as liking moisture and cooler temperatures. But olive trees are not like most of our North American trees. The hotter and drier the weather is, the better they like it.

Olive trees are also very resistant to insects, disease, and decay. Of all fruit trees, which is where olives are classified though their fruit is not sweet, olives are sprayed the least with pesticides and insecticides. They are so naturally resistant that not much outside effort is necessary. Insects seem not even to like the olive tree.

A final interesting and astonishing fact is that eighty to ninety percent of all olives are turned into oil. This was the case in Bible days, and it still is today. This is an remarkably high percentage, and it reveals that the main purpose for growing olive trees is not to get fruit, but oil. It also shows that a great quantity of olives is necessary to produce oil.

We are now ready to look at individual passages that mention the olive. Its first appearance in the Bible is in Genesis 8. In this passage, as the waters of the Flood receded, Noah sent out a raven and a dove to see if the earth was yet dried. The dove, which needed solid ground and a place to rest more than the raven, returned to the ark, indicating that the waters were not yet dried.

So Noah *"stayed yet other seven days; and again he sent forth the dove out of the ark." (Genesis 8:10)* It is interesting that Noah waited seven days. Perhaps this is an indication that the dove was sent out both times on the sabbath day.

This time the dove did not return empty. *"And the dove came in to him in the evening; and, lo, in her mouth was an olive leaf pluckt off: so Noah knew that the waters were abated from off the earth." (Genesis 8:11)*

The olive leaf was in Bible times, and still is today, a sign of peace. When two men or even two large armies are fighting, one eventually extends to the other the "olive branch," requesting peace. One may be winning the contest, or the two may be equal, but one has grown tired of fighting. He then extends the olive branch, indicating that it is time to make peace.

The olive leaf in the dove's mouth was the first extension of this sign of peace. God was telling Noah that the time of destruction was over, and that now it was a time of peace. The sinful men, at war with God, had been destroyed. And Noah now knew, by this olive leaf, that the world was now at peace with

God. God would never again send such a cataclysm of mass destruction by water.

Many people may not realize that this symbol is a part of our daily lives in the United States. Every time we use a one dollar bill, we are holding this symbol in our hands. For if we look closely at the back of our one dollar bill, we will see that the eagle on it is holding an olive branch in one talon. In the other talon, he is holding the symbol of war, a cluster of arrows. These two symbols are placed side by side in order to reflect our national stance. We want peace, the olive branch, but we are prepared for war, the arrow, if it becomes necessary.

The olive branch as a symbol of peace is deeply rooted in history. In Roman times, when Rome fought against and conquered another tribe or country, that conquered people would literally give an olive branch to the Roman captain who conquered them. It was their version of our white flag, their way of saying, "We surrender."

The olive branch was not chosen arbitrarily as a symbol of peace. Rather it was chosen for a specific reason. Wars are very hard on olive trees. When they are destroyed by a war, it takes about eight to ten years for them to grow back and begin to produce olives again. Since their branches have been destroyed, they must grow back from their roots. This is like starting over and takes a good while, and then it takes even longer for them actually to produce fruit again.

In light of this, the olive branch is a symbol of a desire for long peace. It expresses a desire to cease fighting and have a peace long enough for the destroyed olive trees to grow again and bear fruit. Thus to extend the olive branch after a war is to say, "We've had a long war, now let's have a long peace."

It is this symbolism that we find in the dove's olive leaf. In this first appearance of the olive in the Bible, God was expressing His promise of long peace after the long destruction of the Flood.

Another significant mention of the olive is in Deuteronomy 8. Here the children of Israel were about to enter the promised land. They were told, *"For the LORD thy God bringeth thee into a good land, a land of brooks of water, of fountains and depths that spring out of valleys and hills; A land of wheat, and barley, and vines, and fig trees, and pomegranates; a land of oil olive, and honey." (Deuteronomy 8:7-8)*

This bountiful prosperity was given to Israel as a gift. They did not grow the olive trees. They did not plant the wheat. They did not labor for the honey. They did not do anything to bring about this prosperity. Therefore these verses were given to them as a reminder of Who had done all these things for them, *"the LORD thy God."* They were not to forget that it was the Lord Who had provided all these things.

I like to read this verse at Thanksgiving, and even at Christmas when exchanging gifts. It reminds me that I don't do anything to get my gift. It is given to me freely by the other person, just as God blessed Israel and gave them a bountiful land of olives which they didn't do any work to earn. It also reminds me that gifts come ultimately, not even from the human hands that give them, but from God Himself. God is the one we should praise for all gifts and for all things that He freely provides for us .

Another mention of the olive tree is found in one of the most loving and thoughtful commandments in all the law of Moses. Here the Lord commands, *"When thou beatest thine olive tree, thou shalt not go over the boughs again: it shall be for the stranger, for the fatherless, and for the widow." (Deuteronomy 24:20)*

Some today misuse this commandment. They take it as license for the government to take from the rich and give to the poor. But a proper reading of it will reveal that it says no such thing. There are some key words that must be noticed.

First of all, it is a loving admonition for the farmer not to be greedy. No greedy person is truly happy. He always wants just a little more. This commandment is a warning and protection against such an unhappy life. It is given for the farmer's good.

Second, the olives are not provided blanketly for all poor people. In fact, the commandment never even uses the word "poor." The olives are provided instead for three very select and specific people, the stranger, the fatherless, and the widow.

Third, the olives are not given to these people for free. The verse does not tell the farmer to take the olives off the tree and give them away. Instead it tells him to leave them on the tree. The stranger, fatherless, and widow must go and get them themselves. They must do some work, and not just be handed "welfare" for free.

Our modern, un-Biblical welfare system goes entirely against this principle. Yet the system itself is a result of our failure as Christians to fulfill our God-given responsibility to care for the stranger, fatherless, and widow. This responsibility has been given to the church. But the church as a whole has shirked this duty. In the vacuum, the government has stepped in and picked up the responsibility through its un-Biblical welfare program. And as a general rule, what the government picks up, it doesn't give back.

Fourth and finally, the Bible never calls the farmer in this law "rich." He is just a landowner. This law applies to all landowners with olive trees, not just to the rich. And the verses before and after it reveal that the law applies, not just to those with olive trees, but to those with vineyards and fields of any kind.

When we truly read this verse for what it says, there is no way we can use it to justify the government taking from the rich and giving to the poor. Instead, it teaches the opposite. It teaches the value of honest labor for all classes of people, both poor and rich.

We see an excellent example of this law in practice in the lives of Boaz and Ruth. Ruth was both a widow and a stranger in the land of Israel. Boaz, according to the law, left some handfuls of barley for her in the field. But Ruth still had to do the work of gathering the barley herself.

This is a beautiful illustration of the New Testament principle, *"If any would not work, neither should he eat." (II Thessalonians 3:10)* We must not be lazy, but be willing to work for our food.

This principle can be applied spiritually when we see that, throughout the Bible, the olive represents fruitfulness. So if we want to be fruitful, we must work for it. We can't expect our lives to be spiritually fruitful without any effort on our part. God has left us plenty of "olives" on His "tree," but they are left there only for those who will put forth the effort and labor to harvest them. Only those will bear spiritual fruit in their lives.

Another mention of the olive tree is found in Judges 15. Here Samson *"went and caught three hundred foxes, and took firebrands, and turned tail to tail, and put a firebrand in the midst between two tails. And when he had set the brands on fire, he let them go into the standing corn of the Philistines, and burnt up both the shocks, and also the standing corn, with the vineyards and olives." (Judges 15:4-5)*

This was more than just a minor nuisance to the Philistines. To burn up their olive trees was to destroy their very livelihood. It was to destroy their fruitfulness and their entire future, for all of this is wrapped up in the symbolism of the olive tree.

The olive is also mentioned in I Kings 6. This passage describes Solomon's building of the temple, and shows two uses to which he put olive wood. First was to make two cherubims. *"And within the oracle he made two cherubims of olive tree, each ten cubits high." (I Kings 6:23)*

Interestingly, the Hebrew term translated "olive tree" here is again the term *ets shemen* (שֶׁמֶן עֵץ), indicating cultivation. I believe this indicates a holy use of the tree, not necessarily a different kind of tree. Cherubims are the highest created beings in the universe, and these cherubims were in the holiest place in the world, the temple. For such a holy use, God specially cultivated this olive wood through Solomon. This was the very best of the best.

Even in the second use of the olive in the temple, though it was not this specially cultivated olive wood that was used, neither was it the wild olive. There was nothing worldly, wild, or unruly about the temple, but it was entirely holy and sanctified. This is in contrast to the tabernacle. The tabernacle, God's temporary place of residence, was made of wood from the very wild shittah tree. But the temple, God's permanent house, was made of the more refined wood, olive wood.

The second use of the olive in the temple was to make doors and door posts. *"And for the entering of the oracle he made doors of olive tree: the lintel and side posts were a fifth part of the wall. The two doors also were of olive tree; and he carved upon them carvings of cherubims and palm trees and open flowers, and overlaid them with gold, and spread gold upon the cherubims, and upon the palm trees. So also made he for the door of the temple posts of olive tree, a fourth part of the wall." (I Kings 6:31-33)*

It is interesting that cherubims are involved both times olive wood is used. The first time, they are three-dimensional figures, the second time, carvings on the doors. To make these highest created beings, Solomon chose the very best wood of all those used in the temple, olive wood. Olive wood was best because of its resistance to insects, rot, and decay. This quality makes it a very durable, and also very valuable, wood. Thus the best wood was paired with the highest created being.

This also symbolically demonstrates God's perfection in spiritual creation. Though God's physical creation is under the curse of sin and must be subject to decay and disease, God's spiritual creation is above all these. The new creature that He spiritually creates is perfect and cannot be destroyed by the decays and cares of the world. Like olive wood, it will not rot or decay and cannot be affected by "insects."

The olive is not a large tree. It does not yield large slabs of wood, but only small pieces. Maybe these cherubims and doors were made of olive branches and then overlaid with gold to make a solid shape. We do know that they were all overlaid with gold. The wood would then have formed the main structure, and the gold would have filled in the gaps.

We are not told exactly what the cherubims looked like. Cherubims are described in several ways in the Bible, sometimes like men, sometimes like animals. We don't know and probably cannot know their exact form until we reach heaven.

But whatever they looked like, these olive branches would have formed the basic shape, and the outer covering of gold would have provided the finishing detail. The same would have been true with the doors and doorposts. Their basic shape would have been formed by the olive branches, the cherubim figures would have been carved into the wood, and then gold would have overlaid it all. This gold added the symbol of purity to the high quality of the olive wood, linking these cherubims both with high exaltation and with purity.

Another Biblical use of the olive tree was as oil. In Exodus, when God gave Moses the instructions for building the tabernacle, He said, *"And thou shalt command the children of Israel, that they bring thee pure oil olive beaten for the light, to cause the lamp to burn always." (Exodus 27:20)*

This time it was not the wood that was used, but the oil. And it was not just any olive oil, but **pure** olive oil. Pure olive oil is obtained by just bruising the olive, rather than crushing it. If the olive is crushed, it yields pulp mixed with the oil. But if it is just barely squeezed with the fingers and then hung upside down to drip, it is pure oil that comes from the olive.

This is a beautiful picture of Christ Who was *"bruised for our iniquities." (Isaiah 53:5)* He was not crushed, but bruised, and therefore He yielded pure oil.

Pure oil also burns more brightly than oil with pulp. Again, this is a picture of Christ and His Gospel. He, Who was bruised to yield pure oil, is the bright *"light of the world." (John 8:12)*

A personal application can be drawn from this pure oil of the tabernacle when we understand that it had to be constantly refilled by the priests. It was not Israelites in general that refilled this oil, but the priests. The people were to bring in the pure oil, but the priests had the responsibility to refill the lamps and keep them burning brightly.

In the same way today, pastors and those in spiritual authority have the responsibility to keep the light of Christ burning brightly in the world. The only way to do this is to keep the oil, the Gospel of Christ, pure, for it is pure oil that burns the brightest. A lot of pastors today have "watered down" the oil, the Gospel. They have fallen down on the job. But the true pastors don't water down the Gospel. They continually refill their people with the pure and undiluted Gospel, and thereby they keep the light of the Gospel of Christ burning as brightly as a lamp of pure olive oil.

An interesting symbolic mention of the olive tree is found in Psalm 52. Here David said, *"But I am like a green olive tree in*

the house of God: I trust in the mercy of God for ever and ever."
(Psalms 52:8)

The olive tree in this verse is specified as a **green** olive tree. This is to emphasize its youth and therefore its productivity. Even a dead olive tree can still be called an olive tree. But such is not the case with this olive. It is green, which means that it is young and flourishing, and therefore that it produces much fruit.

We see this even with people. Young people have more energy and can produce a lot more fruit than older people. Of course, this fruit can be either good or bad, just as it can with an olive tree. But for this green olive in the house of God, the fruit is good fruit. It spends its youth and energy in the house of God, and therefore produces good fruit.

This verse provides a good contrast with Psalm 37:35, which says, *"I have seen the wicked in great power, and spreading himself like a green bay tree." (Psalm 37:35)* In these two verses, the wicked man and the fruitful Christian are contrasted. The wicked is pictured as a green bay tree, while the fruitful Christian is pictured as a green olive tree.

Like the olive tree, the bay tree is green. It is young, flourishing, and productive. But it is bay tree, not an olive tree, and a bay tree produces no fruit. Thus the wicked, though he may seem to prosper, produces no fruit.

But the green olive tree does produce fruit. It is this true prosperity for which we every Christian should strive. Let us all obtain our productivity from the *"house of God"* and be like the fruitful olive, not the barren bay.

Another place where the olive is used to picture people is in Psalm 128. This entire Psalm describes the blessedness of the man who fears the Lord and walks in His ways. Verse three describes his family. *"Thy wife shall be as a fruitful vine by the sides of thine house: thy children like olive plants round about thy table." (Psalm 128:3)*

This is a happy picture drawn from the characteristics of the olive tree. Young olive plants sprout up from the roots of mature olive trees. So all around a mature olive tree, there are little sprouts coming up. Sometimes such sprouts are called suckers, but usually suckers are actually on the tree itself, not the root.

This is God's picture of blessing. All around him are his children growing up from his roots. This is contrary to many modern notions about children. Today many couples choose to have no children or to have few children, and some even get abortions. This is a tragedy. They are refusing one of God's greatest blessings, that of children growing up about them like young olive plants.

This growth of the young olives continues after the parent dies. Even when the original olive tree is dead, the young olives still grow from its roots and eventually produce their own roots and become trees themselves. The same is true for the family of the blessed man. His children, about his table and home, grow up under his care. When he dies, they continue to grow, and they take over where he has left off.

Also, like young olive plants, children become like their parents. There is a common saying, "The apple doesn't fall far from the tree." In this context, we could say, "The olive doesn't grow far from the tree." In both trees and people, the second generation greatly resembles the first.

There is a group of aspen trees in the Wasatch mountains of Utah that are a good illustration of this. This group contains 47,000 aspen trees that cover an area of about 106 acres. Incredibly, all these trees change color and drop their leaves at the same time. It has been discovered that this is because they all come from the root of one original aspen tree, and so they all are interconnected by their roots. Therefore they are all like that first original parent tree, and really are, in fact, one single tree.

Of course, the second generation is not exactly the same as the first. No two trees and no two people are exactly the same. But sons and daughters are similar to their parents in looks, characteristics, and actions. They often even laugh like their parents. We have all no doubt had the experience of meeting the son or daughter of someone we know. It is usually pretty easy to tell who the parent is, because of these similarities.

These olive plants around the blessed man's table are also a sign of fruitfulness, fatness, and richness. Fruitfulness is God's desire for each life, and here in this home, we find this desire fulfilled. The children of this kind of man, a man who fears the Lord and walks in His ways, are indeed fruitful.

In the book of Job, the olive appears in the opposite context, mentioned by one of Job's "comforters," Eliphaz. The overall objective of Eliphaz and his friends was supposed to be to comfort Job, but in this specific passage, Eliphaz was not trying to comfort Job at all. He was trying to accuse him.

Eliphaz said, *"Let not him that is deceived trust in vanity: for vanity shall be his recompence. It shall be accomplished before his time, and his branch shall not be green. He shall shake off his unripe grape as the vine, and shall cast off his flower as the olive. For the congregation of hypocrites shall be desolate, and fire shall consume the tabernacles of bribery. They conceive mischief, and bring forth vanity, and their belly prepareth deceit."* (Job 15:31-35)

Here Eliphaz was telling Job that wicked men were miserable. He could see that Job was miserable, and so he tried to conclude from this that Job was a wicked man. But his logic was flawed.

Eliphaz's hypothetical statement was true, "If you are wicked, then you are miserable." His second statement was also true, "Job is miserable." But his conclusion was false, "Job is wicked." This is because he had a faulty logic structure.

This faulty structure can be easily seen when we change the terms, but keep the structure. Our first statement becomes, "If you are a dog, then you have four legs." Our second statement becomes, "A cat has four legs." Both of these statements are true. But if we use Eliphaz's logic, we reach an absurdly false conclusion, "A cat is a dog."

Just because Job was miserable did not automatically mean that he was wicked. Wicked people truly are miserable, but not all miserable people are wicked. Misery can be brought about by causes other than wickedness.

But Eliphaz persisted on in his argument. He used the olive to picture the untimely death of Job's children. He said, *"He shall . . . cast off his flower as the olive."* (Job 15:33) This tragedy was described in chapter 1. *"And, behold, there came a great wind from the wilderness, and smote the four corners of the house, and it fell upon the young men, and they are dead; and I only am escaped alone to tell thee." (Job 1:19)*

We see here that it was a great wind that killed Job's children. This ties in very interestingly to our study of the olive. Greeks are the number one growers of olives. But they periodically have a severe problem that hurts their crop. Occasionally a very strong wind from the north sweeps in over the mountains. This wind is so strong that it blows the flowers off the olive trees. It has become known as the Greco-Tramontane, Greco meaning "Greek" and Tramontane meaning "over the mountains."

This Greco-Tramontane causes great harm and loss to the olive growers of Greece. It comes during the springtime when the trees are the most tender and have only blossoms. But these very blossoms are the parts that would eventually become the fruit. So when they are blown off the trees in the spring, the trees produce less, or even no, fruit in the harvest.

It was this imagery that Eliphaz used to describe the death of Job's children. Job's young and tender olive blossoms were blown off his tree by a strong wind. In his home, they would no longer sit like olive plants round about his table.

So what was Job going to do? It seemed perhaps that Eliphaz was right. Job had not brought forth fruitful olive trees. It seemed that he was not the blessed man of Psalm 128, and that he was being judged because he was wicked.

But God had other words to speak on the subject in the book of Haggai. *"Is the seed yet in the barn? yea, as yet the vine, and the fig tree, and the pomegranate, and the olive tree, hath not brought forth: from this day will I bless you." (Haggai 2:19)*

Here in Haggai, the circumstances looked just as bleak as they did for Job. There was no fruit in the field. But God asserted that the circumstances were not bleak. Though no fruit could be seen, yet the barn still had seed. Contrary to all apparent circumstances, God promised, *"from this day will I bless you." (Haggai 2:19)*

In Job's case, the same promise applied. He hadn't brought forth fruit yet, and it looked like all hope of fruitfulness was gone. The flowers were all blown off. Yet in spite of this seeming hopelessness, God was going to bless him and give him fruit miraculously.

God promises the same for us. Though our circumstances may look bleak and hopeless, He will bless us later with fruit if we will but trust Him. We cannot trust ones like Eliphaz who say that we are miserable only because we are wicked. Of course, there are times when we are miserable because of our wickedness. But there are other times when we are miserable, but not because of our wickedness. At these times, trust Christ. He will put the blossoms and eventually the olives back on our trees.

Another interesting mention of the olive tree is in Isaiah 17. Here the Lord is describing the coming judgment upon Israel. He tells of all the people that will be carried away by the enemy, Sennacherib. But then He gives this reassuring promise, *"Yet gleaning grapes shall be left in it, as the shaking of an olive tree, two or three berries in the top of the uppermost bough, four or five in the outmost fruitful branches thereof, saith the LORD God of Israel." (Isaiah 17:6)*

Not all the people will be carried away. When an olive tree is shaken, which it must be in order for it to release its fruit, most of the olives fall to the ground, but not all. In the top of the tree, a few fruits are left. We can see this when we shake a nut tree or even a stalk of corn. Always there are a few nuts or corn kernels left in the top. Even when fruits are gathered by hand, not every one is harvested. A few are always left.

This is because the top fruits of any plant are further from the sap than the bottom fruits. For the sap to reach the top fruits and bring them nutrients, it must work against gravity. Thus the bottom fruits ripen earlier. So when a tree is shaken, it is the ripe fruits on the bottom that fall off, while the unripe ones on the top remain on the tree. Here in Isaiah, these remaining fruits picture the preserved remnant that is left in the land.

The scientific name for the very top of a tree is "amir," from which we get our word "admiral," the title of the "top man" in the navy. Just as it is this top part that is spared on the olive tree, so it is the top of Israel that is spared from judgment. Jerusalem is the chief focal point of the land of Israel. It sits high in the mountains in a very fortified location. Consequently, it is usually the very last place that an enemy army conquers. It is the amir, the top, the spared remnant.

Here in Isaiah, Sennacherib and his armies try to destroy Israel. They do their very best to bring total destruction. But God makes it so that they cannot find everyone and everything. He thereby mercifully leaves a remnant.

This is the way God works with His people. Though He may have to bring judgment, yet He always leaves a remnant. It is not this way for wicked nations like the Amorites. These He destroys completely. But for His Own, He leaves a remnant.

In the last days, we are told that all nations will come against Israel. Yet this loving God will, as always, leave a remnant. He will cause some to escape into the wilderness, and thereby preserve a remnant of His Own people.

Later in Isaiah, this remnant is spoken of again. This passage shows the great joy that will come to these spared people. *"When thus it shall be in the midst of the land among the people, there shall be as the shaking of an olive tree, and as the gleaning grapes when the vintage is done. They shall lift up their voice, they shall sing for the majesty of the LORD, they shall cry aloud from the sea." (Isaiah 24:13-14)*

This principle of God sparing a remnant appears over and over in Scripture. In the wicked days of Ahab and Jezebel when it seemed that Israel was given over to idolatry, God told Elijah, *"Yet I have left me seven thousand in Israel, all the knees which have not bowed unto Baal, and every mouth which hath not kissed him." (I Kings 19:18)*

In the days of Noah, when *"the wickedness of man was great in the earth, and . . . every imagination of the thoughts of his heart was only evil continually," (Genesis 6:5)* God still had a remnant in Noah and his family. He destroyed all but them.

Even in the cities of Sodom and Gomorrah, God preserved His remnant. He delivered Lot and his two daughters from the fires of judgment.

There is, however, an important contrast between Noah and Lot. Noah wanted to escape from the wickedness around him, but Lot didn't. Lot was comfortable living in the world and had to be dragged forcibly out of it. Noah, on the other hand, was eager for deliverance from the wicked world about him.

God's mercy upon these, however, does not mean that His people are always free from punishment for their sin. On the contrary, as a holy and just God, He must not only punish the wicked, but His Own people as well. Jeremiah uses the olive tree to illustrate this very point.

In Jeremiah's day, the sin of Judah and Jerusalem had become very great. They had turned their backs on Jehovah and had turned to idolatry. For this sin, God pronounced judgment upon them. The judgment was to be so severe that He said, *"Therefore pray not thou for this people, neither lift up a cry or prayer for them: for I will not hear them in the time that they cry unto me for their trouble. What hath my beloved to do in mine house, seeing she hath wrought lewdness with many, and the holy flesh is passed from thee? when thou doest evil, then thou rejoicest." (Jeremiah 11:14-15)*

Jeremiah was not even to pray for this sinful people. God's sentence was pronounced, and their judgment was sealed. There are times when the sin of a people is so great that we are not even to pray for their deliverance from God's just destruction. Such was the case here. When people completely turn their backs on Christ and turn to idolatry or anything else, they must be judged.

The passage goes on to liken Israel's destruction to that of an olive tree. *"The LORD called thy name, A green olive tree, fair, and of goodly fruit: with the noise of a great tumult he hath kindled fire upon it, and the branches of it are broken. For the LORD of hosts, that planted thee, hath pronounced evil against thee, for the evil of the house of Israel and of the house of Judah, which they have done against themselves to provoke me to anger in offering incense unto Baal." (Jeremiah 11:16-17)*

Their sin was great, and they were doomed to be burned up like an olive tree. However, just because we are not to pray for such people to be delivered from destruction, this does not mean that we are not to pray for them at all. We are still to pray for their salvation. We should pray that they will turn from their willful and sinful ways and turn back to Christ. The only thing we should not seek to alter is God's appointed judgment upon their sin.

This weighty passage is given to us so that we may know that sin has a cost to it. Christ cannot just wink at sin and let it pass. If He were to do that, He wouldn't be Christ. Because of His just and holy nature, sin has to be punished.

In this case, Israel had willfully and persistently offered incense to idols. Therefore, according to God's standard, they had to be judged. Using the analogy of the olive tree, they had used their richness and fatness for purposes other than Christ and His glory. Therefore Christ had to judge them.

But surprisingly, it is due to times of judgment that some of the sweetest oil is produced. In fact, some of the best olive trees grow in rocky, dry crevices. This is different from most trees. We mentioned the "oaks of Bashan," which literally means "oaks of soft, fertile soil." Oaks flourish best in soft soil. But the olive tree is different. It grows best in chalky and rocky places.

Thus we see the astonishing and ironic truth that it is out of the dry and rocky places that sweet oil flows. It is from the rocky crevices that we obtain our oil, fatness, and fruitfulness.

We are told that it is in just such a rocky and barren place that the remnant of Israel will be sheltered in the last days. Revelation 12 pictures this remnant of Israel as a woman. When the dragon comes against her, she flees *"into the wilderness, where she hath a place prepared of God, that they should feed her there a thousand two hundred and threescore days." (Revelation 12:6)*

A place of hiding is usually not a place of comfort. It is not a nice house with air conditioning, carpet, servants, and plenty of food. Rather it is an unlikely place of dwelling, in order that the hiding place not be discovered. Yet it is in this austere, rocky place that Israel produces her sweetest fruit. And it is in the rocky places of our lives that we also produce our most bountiful richness and fatness for the Lord.

This fruitfulness is what Christ desires from our lives. But don't we do the same? When we plant a garden, we expect it to bear fruit. If we plant carrots, tomatoes, beets, and cucumbers, but they don't bear fruit, do we leave the plants in our garden? No, we plow them up or mow them down.

In the same way, Christ will not leave us in His garden if we are barren plants, merely taking nutrients from the soil and not bearing any fruit. No, He will root us out and replace us with something that produces fruit, something that is profitable to Him.

We see this illustrated with the olive tree in the book of Amos. The children of Israel had not produced fruit like they were supposed to, and so the Lord told them, *"I have smitten you with blasting and mildew: when your gardens and your vineyards and your fig trees and your olive trees increased, the palmerworm devoured them: yet have ye not returned unto me, saith the LORD. I have sent among you the pestilence after the manner of Egypt: your young men have I slain with the sword, and have taken away your horses; and I have made the stink of your camps to come up unto your nostrils: yet have ye not returned unto me, saith the LORD." (Amos 4:9-10)*

The olive is usually resistant to insects, decay, and disease. Hardly anything can attack it. But here, contrary to nature, the Lord devoured it with the palmerworm.

Just as the olive is usually resistant to insects, so the children of Israel were usually resistant to idolatry. They were God's chosen people and had seen his miracles. They were also usually resistant to calamities, for they were miraculously protected by the Lord. However, when they did turn to idolatry, the Lord caused them to be attacked and judged by trials.

So also we as Christians are usually resistant to idolatry, and also to trials and tribulations. Of course, this is true only of true Christians. Those who fake Christianity are quickly drawn to idolatry and are easily affected by trials and problems in their lives. But as true Christians, we know the power of Christ, and we have Jesus Christ to lean upon through our trials.

However, the temptation to idolatry is very strong today. In fact, we may have more idols today than the children of Israel had in Amos' day. For anything that we put before Christ is an idol. Houses, cars, children, parents, food, and all kinds of things can be idols, if we put them before Christ. Today we don't often

the captivity, Ezekiel was the priest, and Daniel was the lawgiver. Even in Christ, these two distinct offices can be seen, though they are combined into one person. For Christ is both our High Priest and our Lawgiver.

These same two offices seem to be applied to the two witnesses in Revelation. One witness is the priest, and the other is the lawgiver. But they are not Joshua, Zerubbabel, Ezekiel, or Daniel. So there comes the rub. Who are these two witnesses? One, I think, is definitely identifiable as Elijah. Instead of dying a natural death, he was caught up to heaven in a fiery chariot, and his body seems to have been taken to heaven to reappear again on earth at this time. For the identity of the other witness, the two most debated are Moses and Enoch.

Whatever their exact identity, by spreading the pure Gospel of Christ, these two witnesses, like the olive trees of Zechariah, will give oil to the candlesticks of the Gospel. Indeed, prophesying at the end of the *"times of the Gentiles," (Luke 21:24)* with the rapture having already occurred, they will be the only supply of this pure oil and of Gospel light in the world. Their candlesticks will burn all the more brightly and thereby lighten the dark world with the bright light of the Gospel of Jesus Christ.

We now move on to some other mentions of the olive in the New Testament, beginning with the one in the book of James. James uses the olive to teach the importance of having a sweet tongue.

James says of the tongue, *"But the tongue can no man tame; it is an unruly evil, full of deadly poison. Therewith bless we God, even the Father; and therewith curse we men, which are made after the similitude of God. Out of the same mouth proceedeth blessing and cursing. My brethren, these things ought not so to be." (James 3:8-10)*

He then goes on to make some analogies. *"Doth a fountain send forth at the same place sweet water and bitter? Can the fig tree, my brethren, bear olive berries? either a vine, figs? so can no fountain both yield salt water and fresh." (James 3:11-12)*

Most olives are not sweet, but salty or even bitter, especially when they are green. They don't taste anything like a sweet fig.

James is teaching here that the tongue should yield sweet fruit at all times, not sweet fruit at some times and bitter fruit at other times.

I probably shouldn't have done this, but when my daughter was little, I convinced her that a black olive was a grape. She ate it, but instantly, she knew that it was not a grape. The bitter taste was quite a shock to her when she was expecting something sweet. Even in this illustration is an example of how hard it is to tame the tongue. It was my false tongue that made my daughter think that something salty was sweet instead.

Romans 11 uses the olive to illustrate the principle of grafting. It says to us as Gentiles, *"Thou, being a wild olive tree, wert graffed in among them, and with them partakest of the root and fatness of the olive tree." (Romans 11:17)* We covered this passage in chapter 22 of this book, the chapter on grafting. So we won't cover it again here.

The last reference I want to look at is not necessarily to the olive tree itself, but to the Mount of Olives, so named because of the many olive trees that grow there. Zechariah prophesies of Christ's second coming, *"And his feet shall stand in that day upon the mount of Olives, which is before Jerusalem on the east, and the mount of Olives shall cleave in the midst thereof toward the east and toward the west, and there shall be a very great valley; and half of the mountain shall remove toward the north, and half of it toward the south." (Zechariah 14:4)*

This Mount of Olives is also called Olivet. Using this name, Acts tells us that Jesus ascended back to heaven from this very same mountain. *"And when he had spoken these things, while they beheld, he was taken up; and a cloud received him out of their sight. And while they looked stedfastly toward heaven as he went up, behold, two men stood by them in white apparel; Which also said, Ye men of Galilee, why stand ye gazing up into heaven? this same Jesus, which is taken up from you into heaven, shall so come in like manner as ye have seen him go into heaven. Then returned they unto Jerusalem from the mount called Olivet, which is from Jerusalem a sabbath day's journey." (Acts 1:9-12)*

There is much in that phrase *"in like manner."* Zechariah reveals that it includes Christ coming to the very same mountain, the Mount of Olives. It is to this fruitful place, this place that is symbolic of fatness and richness, that Christ will return. Let us be ready for that day.

34
The Palm Tree

The palm tree is a very unusual tree. It is an evergreen in that it doesn't drop its leaves at a specific time of the year. But as a monocot, it really belongs in the grass family, not the tree family.

The palm is also an amazingly varied tree, there being over 2,500 different varieties of palms. Some familiar ones are the sable palm of Florida and the coconut palm. In Israel, the most common palm is the date palm, which is grown in great abundance.

Indeed, because of the abundance of palms in Palestine, it was known in Bible times as the land of palms. Jericho, the first city that the Israelites took in the promised land, was known as *"the city of palm trees." (Deuteronomy 34:3)* Today, due to the urban environment of Israel and the wars in its recent history, there are not as many palms as there used to be. But the date palm is still widely grown for its fruit.

The palm is rich in Biblical symbolism. It is used to represent rest, relaxation, refreshment, and rejuvenation, both of spirit and of body.

In I Samuel 23, we read that David *"went up from thence, and dwelt in strong holds at Engedi." (1 Samuel 23:29)* Engedi is a fresh water spring in the midst of the hot Judean wilderness. Its cool waters make an oasis where many palm trees flourish. The shade, water, and greenery provide a welcome refreshment from the desert environment all around. In this passage, David had just been hotly pursued by Saul. At the last moment, just as Saul was on the other side of a mountain from David, Saul was called away to fight the Philistines who had just invaded the land. David, thus delivered from Saul and weary after many anxious days of flight and hiding, found rest and rejuvenation among the palm trees of Engedi.

Another place where the refreshment of the palm tree is seen is in Exodus 15 during the Israelites' journey from Egypt to Canaan. The children of Israel had just gone three days without water, and then had miraculously been given water by the Lord at Marah. As a welcome refreshment to this difficult stretch of journey, the next place they came to was Elim. *"And they came to Elim, where were twelve wells of water, and threescore and ten palm trees: and they encamped there by the waters." (Exodus 15:27)* Here under the shade of the palm trees, the people found rest and refreshment for their weariness. This teaches us still today that even in the desert, the Lord is faithful to give His people the rest and rejuvenation that they need.

A well-known Biblical account of the palm tree is commemorated every year on Palm Sunday. Interestingly, the entire basis for this name is found in just one verse of the Bible. All the Gospels record the triumphal entry of Jesus into Jerusalem, and both Matthew and Mark record that branches were cut down and strawed in the way of Christ. But only John records that these branches were palm branches. *"On the next day much people that were come to the feast, when they heard that Jesus was coming to Jerusalem, Took branches of palm trees, and went forth to meet him, and cried, Hosanna: Blessed is the King of Israel that cometh in the name of the Lord." (John 12:12-13)*

What kind of palm trees were these? The Bible doesn't say, but we can gather a few clues. We have already noted that the date palm is the predominant palm in Israel, and therefore it is a possibility. Obviously, these palms were probably not coconut palms. Coconut palm branches would have been much too high for the people to reach easily. The passage implies that the branches were within easy reach, and so these were probably date palms with low branches and a bush-like shape.

Why did the people spread these palm branches before Christ? It is because of the rich symbolism of the palm. In the Old Testament, aside from being the symbol of refreshing, the palm is primarily the symbol of rejoicing. In the New Testament, it is primarily the symbol of victory. Here both the Old Testament and New Testament symbols were merged. The Jewish nation had been waiting for their Messiah to come and establish His kingdom and free them from the Romans. As Christ entered Jerusalem on an ass's colt just as had been prophesied, the people thought that He would bring in the kingdom right then. They therefore took up the palm branches to symbolize the victory of Messiah Who had come to free them and simultaneously to symbolize their rejoicing in His victory.

The palm tree as a symbol of victory in New Testament times was seen even in the secular world. The victor of a Greek or Roman race would be given a crown made of a palm branch. A victor returning from war or a returning victorious army would be met with palms spread in the way. Thus by spreading palms in the way of Christ, the people were recognizing Him as the Victor.

The use of the palm tree here brought to the minds of the people their spiritual heritage, as pictured in the temple. The temple, the center of life for the Jewish people, was adorned with images of palm trees. I Kings 6-7, which records the building of Solomon's temple, is abundant with mentions of the palm tree as it was imprinted throughout the temple.

Even the future of the Jewish nation will retain the symbol of the palm tree. The restored millennial temple of Ezekiel 40-41, like the temple of Solomon, will also be filled with images of the palm tree. Ezekiel prophesied of this temple, *"And it was made with cherubims and palm trees, so that a palm tree was between a cherub and a cherub; and every cherub had two faces; So that the face of a man was toward the palm tree on the one side, and the face of a young lion toward the palm tree on the other side: it was made through all the house round about. From the ground unto above the door were cherubims and palm trees made, and on the wall of the temple." (Ezekiel 41:18-20)*

This is not referring to actual palm trees anymore than it is referring to real cherubims. Both are images. The wood of the palm tree is not used in the temple because it is not a structurally sound wood. The first wind would blow it down. Instead fir and cedar wood, which can withstand the wind, are used for construction. This wood is then carved with the image of the palm tree and then overlaid with gold, which conforms to the shape of the carved palm tree underneath.

For construction, wood must have a balance of flexibility and strength. It can't be too flexible, or it will fall over. But neither can it be too rigid, or it will break. It needs to be able to flex and give a little, but not too much. The oak and pine are good examples of this balance. They are both flexible and strong.

The palm tree, because it is a monocot and in the grass family, is very flexible, but not strong. Its wood is very fibrous like the stalk of any grass, and its density is too varied to be reliable. Trying to drive a nail into it is like trying to drive a nail into thick rubber. It can't be done. It splits the fibers, and the nail comes out. Thus the palm tree was not used in construction, but its image was imprinted all over the temple. Adorning the walls, doors, posts, and altars, it was a very important symbol for the Jewish people. Even Jewish coins were imprinted with palm trees, and still are today as a symbol of their nation.

The growth of the palm tree is an interesting characteristic. Its trunk doesn't grow outward and get wider like the trunks of most trees, but instead it grows upward only. As the leaves fall, the trunk shoots up further, but it remains the same width all the way up. Some palm tree trunks are as small around as pencils and never get any bigger than that.

Another amazing fact about the palm tree is the ease with which it can be transplanted. When transplanting most trees, for every inch of diameter of the trunk, the tree needs one foot of root ball to be transplanted with it. But the palm tree can be transplanted with much less of a root ball than this. This ability to withstand a lot of root cutting is due to the fact that it is a grass. When it is transplanted, it develops a tap root that goes very deep and keeps the tree from falling.

The lifespan of the palm tree is yet another of its notable characteristics. The coconut palm has a very rigid and fixed lifespan, living and producing fruit for only a short and predictable number of years. But the date palm, the common palm of Israel, has a very different lifespan. It can live a couple of centuries. It is used in the Bible to show the longevity and prosperity of the righteous. *"The righteous shall flourish like the palm tree."* *(Psalm 92:12)* The longevity of this palm demonstrates not only the long physical life that is promised to the righteous, but also the long and full spiritual life of Christ within them.

The palm tree, a grass by technical terms, is very different from all the other trees of the Bible, both in its physical qualities and its spiritual symbolism. It stands as the supreme symbol of Christ's victory. Just as the palm is ever green, so Jesus Christ has always been victorious, is still victorious, and will forever be victorious. Though His kingdom was not established at His triumphal entry, the palm still was rightly spread before Him. For indeed He is the Victor and will gain the ultimate victory.

35
Parts of a Tree

The various parts of a tree form an instructive analogy to the Christian life. The first part of a tree is its **roots**, called in Greek by the word *rhiza* (ριζα). In the Bible, the roots represent of Jesus Christ. Isaiah prophesies of Him, *"And there shall come forth a rod out of the stem of Jesse, and a Branch shall grow out of his roots." (Isaiah 11:1)* Again Isaiah says of Christ in chapter 53, *"For he shall grow up before him as a tender plant, and as a root out of a dry ground: he hath no form nor comeliness; and when we shall see him, there is no beauty that we should desire him." (Isaiah 53:2)*

Like the root of a tree, Jesus Christ gives vital sustenance to all the rest. Without the root, the tree would be lost. For it cannot be sustained without the nourishment provided by the root.

Another picture of the root is found in the book of Malachi, where the Lord pronounces this judgment upon the wicked among His people, *"For, behold, the day cometh, that shall burn as an oven; and all the proud, yea, and all that do wickedly, shall be stubble: and the day that cometh shall burn them up, saith the LORD of hosts, that it shall leave them neither root nor branch." (Malachi 4:1)* By saying that He will leave them neither root nor branch, the Lord is saying that He will leave them neither a present life nor a future life. The removal of the branch will cut off their present life, but the removal of the root will cut off their future life. This is complete judgment upon these wicked men.

An interesting analogy to the Christian life can be drawn from the growing habits of roots. Often when one buys a tree from a nursery to plant at home, the nursery plant has lived its entire life in very nutrient-rich soil. When it is put in the new soil, it finds that the new soil doesn't have the same amount of nutrients. When the roots detect this, they begin to grow in a circle, staying right around the nutritious soil that is just at the base of the tree. This condition is called girdling roots. If the roots continue to grow this way, the tree will eventually choke itself to death.

For this reason, adversity must be sent to encourage the roots to grow out from the fertilized soil and out into the barren soil. This is also the reason that the Lord sends adversity into our lives. We are not used to barren soil and so are reluctant to venture into it. But the adversity, lovingly sent by Christ, drives us there for our good.

Adversity, as distressing as it is, is always sent for our good. A wise nurseryman will walk through his plants from time to time and shake them. This gives them some adversity by simulating the wind blowing them and causes the cells to grow stronger. Then when they are removed from the nursery and planted in a harsher environment, they can stand through strong wind without falling over or breaking off.

The consequences of not enduring such adversity can be seen when the trees on the outer edge of a forested area are cut down. The trees in the center are then for the first time in their lives exposed to strong wind and increased adversity. Having not had early adversity to strengthen them, their branches easily break.

Christ knows just the right amount of adversity to bring into our lives, not too little that would make us weak, and not too much that would cause us to break. Christ has the perfect balance of adversity and protection.

In every plant, there is a root-shoot balance. If part of the roots die, a proportionate amount of the top shoots die. And if part of the top shoots die, a proportionate amount of the roots die. We as Christians must have a balance as well, a balance between our root strength and protection, and our shoot vulnerability to adversity. This balance Christ ensures for us, by controlling the amount of adversity and fertilizer that He gives us.

The **stem and branches** are the next part of the tree. They represent people, either attached to the root of Christ or not attached, either fruitful or unfruitful. We have covered the branches in great detail in chapters 19 and 22 on fruit and grafting, and we will cover them again in chapter 39 on pruning. We will also cover Christ as the branch in chapter 47 on the tree of life.

The **leaves** are another part of the tree. The Greek word for them is *phullon* (φυλλον). They represent the works and the outward actions of one's life. The leaves are an important part of the tree. While they are not themselves fruit, the outward actions of a person do reveal his spiritual health and therefore his potential to bear fruit.

But perhaps one of the most vital yet overlooked parts of the tree is the **seed**, called in Greek *sporos* (σπορος). The seed is the smallest part of the tree, yet it is the key to the tree's very life.

John 12 presents us with a paradox on the seed. The seed, the key to life, must first die. *"Verily, verily, I say unto you, Except a corn of wheat fall into the ground and die, it abideth alone: but if it die, it bringeth forth much fruit." (John 12:24)* Only through death can the seed give life and bring forth fruit.

Death in the Bible is not annihilation, but separation. At a person's death, they do not cease to exist like an animal does. Instead their soul is separated from their body. At this time of separation, the bodies remain on the earth, while the unsaved souls go to Hell and the saved souls to Heaven.

The death of the seed is also separation. Every seed has two parts, the shoot apical meristem and the root apical meristem. When the seed dies, these two separate one from another in order to form the root and stem of the new plant.

Seeds also have three kinds of cells, the parenchyma cell, the collenchyma cell, and the sclerenchyma cell. These picture the three parts of a person's life, the body, the soul, and the spirit. After death, a person receives a new body, and his soul and spirit live eternally in that body. In the same way, the outer cell of the seed, picturing the body, breaks open upon the death of the seed, and a new outer structure, the new plant, forms. The two other cells, picturing the soul and the spirit, live on and are the very life of the new plant.

In order to germinate, the seed must undergo two processes, known as scarification and stratification. Scarification is the breaking of the external seed coat, and stratification is the breaking of the internal seed coat. Both of these are necessary, or the seed will not germinate.

In the same way, a person who is genuinely saved will have both an internal change and an external change. If a person says that he is saved and has been changed on the inside, but there is no evident change on the outside, then perhaps there has been no true change on the inside. A change on the inside will always manifest itself on the outside.

On the other hand, a change on the outside does not necessarily mean there has been a change on the inside. A person may simply reform himself. If this is the case, only the outside is changed, and there has been no change on the inside. There must be both. Only when there is a true change on the inside will there be a true and lasting change on the outside as well.

This is the death that the seed must undergo, the separation of its seed coats and of its cells from one another. It is only through this death that one may receive new life, for as I Corinthians 15 restates, *"that which thou sowest is not quickened, except it die." (I Corinthians 15:36)* Only through death can a seed, or a Christian, have new life. Only through this death, this separation, can we bring forth fruit.

This same passage in I Corinthians also speaks of the coming day when all such separation will be forever over. *"The last enemy that shall be destroyed is death." (I Corinthians 15:26)* Understanding death as separation, this verse literally says that the last enemy that shall be destroyed is separation. In heaven, there will be no more separation. There will be no more separation from our loved ones, our friends, and our family.

Yet before we reach this blessed state, there is yet one more kind of seed hidden deep in our hearts. This kind of seed resides in all hearts. It is the seed of sin. Every person has these seeds of sinning. It only takes certain conditions for these seeds to grow.

These seeds we must choke out. We must not allow the conditions that draw them out of us. We must not cultivate them and allow them to take root in our lives.

While we live on earth, let us submit to the separation the Lord has for us, the breaking and transforming of our lives both internally and externally. Through this death only may we draw upon Christ the root, grow into branches, manifest leaves, and eventually bring forth fruit.

36
The Pine Tree

The pine tree is mentioned only three times in the Bible. Though the word "pine" occurs more often, this is because it is also used as a verb, as in the phrase "pine away." But this is an act that is done, not a substance like a tree. For the actual pine tree, there are only three times it is mentioned.

The Hebrew word for the pine tree is *tidhar* (תדהר). Various commentaries offer many suggestions as to what this tree is. The list includes elm, oak, juniper, cypress, cedar, and holly. They are all over the map when it comes to this tree. But there are ways to eliminate some of these.

When this tree is mentioned in Isaiah, both the other trees listed with it are evergreens. This is a key. It seems that this tree also must be an evergreen. This rules out the elm and the oak. Though there is one oak in the southern United States, the live oak, that is an evergreen, usually the oak is not an evergreen.

The juniper, the cedar, and the cypress are all already mentioned elsewhere in the Bible, so why would they be called by two different names? Also their cones are different from the cones of the pine tree. The holly tree is not mentioned in the Bible, but its Latin name is *Ilex*, different from that of the pine, *Pinus*.

The specific names of trees in the Bible are given for specific purposes, just as the names of people and places are. They all have specific meanings. Otherwise the Bible would just say "tree," rather than specifying if it is a fig, an oak, or a pine. So I believe that the name of this tree as pine is given in the Bible for a specific purpose.

Probably it is most closely related to the fir tree in the Bible, which is in the pine family. But these two are definitely not the same tree. Twice in the book of Isaiah, the fir and the pine appear together in lists of various trees in the same verse. They therefore cannot be the same tree, for this wouldn't fit with the variety that is intended in these verses.

There is a pine known as the Jerusalem pine because it grows abundantly around Jerusalem. This is the Aleppo pine. This may be the Biblical pine, or it may not be. Of course, we know that the land of Israel is different today than it was in Biblical times. It is dry and desolate today. While a few pines will grow in such a climate, it is not what they prefer. So there could have been more pines and different pines in Israel in Biblical times than there are now.

The pine cone is rich in symbolism. It is the same shape as the top of the human spinal cord and as the cones that are embedded in the human retina. It is also the same shape as that used by the Catholic church for the top of the pope's staff, the top of the Vatican flag staff, and the pope's crown. Interestingly the largest sculpture of a pine cone is found in the Vatican. This is because the pine, as an evergreen, is a symbol of everlasting life. Thus its cone is a symbol of the fruit of everlasting life. So whoever owns it supposedly has the fruit of everlasting life.

The first Biblical mention of the pine tree is in Nehemiah 8. It says of the children of Israel reentering the land of Israel after their captivity, *"And they found written in the law which the LORD had commanded by Moses, that the children of Israel should dwell in booths in the feast of the seventh month: And that they should publish and proclaim in all their cities, and in Jerusalem, saying, Go forth unto the mount, and fetch olive branches, and pine branches, and myrtle branches, and palm branches, and branches of thick trees, to make booths, as it is written. So the people went forth, and brought them, and made themselves booths, every one upon the roof of his house, and in their courts, and in the courts of the house of God, and in the street of the water gate, and in the street of the gate of Ephraim."* (Nehemiah 8:15-16)

In this passage, the children of Israel had just reentered the land of Israel after being in captivity for seventy years. Before their captivity, they had celebrated the feast of tabernacles, but they had not celebrated it during the years of captivity. Now back in the land, they found that they were to observe the feast of tabernacles, and thus they celebrated it for the first time in

there, we can know that God has planted it. In love, He plants for us gardens in the deserts of our lives.

The last mention of the pine tree is also in Isaiah, in chapter 60. It is a fitting close because it speaks of the last days of the earth, the millennial kingdom when the temple will be restored. It says, *"For the nation and kingdom that will not serve thee shall perish; yea, those nations shall be utterly wasted. The glory of Lebanon shall come unto thee, the fir tree, the pine tree, and the box together, to beautify the place of my sanctuary; and I will make the place of my feet glorious." (Isaiah 60:12-13)*

The glory of Lebanon is none other than her cedars. These magnificent trees are her crown of glory. All over the world, she is known for these tall, strong, majestic trees. And in this day, these trees will bring all of their magnificence to the restored temple.

Along with these cedars, the fir, pine, and box tree will each bring their own beauty, and together they will all glorify and beautify the temple. The fir, the pine, and the box, though not as famous as the cedar, each have their own unique contributions to bring. They are all different and therefore all necessary to the complete beauty of the temple. If the pine tree, or even the small box shrub, were to be left out, the temple would be lacking some of its full beauty.

For this reason, none will be left out. All these trees will be dug up by their roots and transplanted into the temple. They will there put down their roots once again and each beautify the sanctuary in their own unique way.

These trees are a picture of the nations spoken of in verse 12. All the nations will come to this millennial kingdom, and those that don't come will perish. The diversity of trees pictures the diversity of the nations. Each nation, like each tree, is unique and has its own unique contributions to bring.

But the trees are not only diverse. Their diversity is overarched by a larger unity. For while they are each different, they are also all part of one whole in that they are all evergreens. In the same way, the nations of this day will all share one thing in common. They will all be everlasting. Joining themselves to the kingdom of heaven, they will become part of the kingdom that will never perish, the "evergreen," everlasting kingdom. Of this coming, glorious day, the evergreen pine is a constant reminder.

37
The Pomegranate Tree

The word "pomegranate" is found several times in the Bible, referring sometimes to the tree itself and sometimes to the fruit. The highest concentration of references to the pomegranate is found in the Song of Solomon.

The Hebrew name for the pomegranate is *rimmown* (רמון). Its English name is a compound. Its first part comes from the Latin *pomum*, meaning apple. Of course, the pomegranate is not a true apple, but an apple in the sense of a round, fruit-like thing. The second part of the word "pomegranate" comes from the Latin *grenat*, which means "red," or "dark red." So the word "pomegranate" literally means "red apple." Some say that the word *grenat* means "many seeds." In this case, "pomegranate" would mean "many seeded apple."

Indeed the most notable thing about the pomegranate is its great number of seeds. It supposedly has 613 seeds. This is a significant number because there are 613 laws in the Old Testament. I'm sure, however, that the number of seeds in a pomegranate depends on the weather.

The Biblical use for which the pomegranate is best known is for lining the hem of the high priest's robe. It was lined alternately with *"a golden bell and a pomegranate, a golden bell and a pomegranate, upon the hem of the robe round about."* *(Exodus 28:34)* Supposedly when we calculate how many pomegranates would have been on one robe, and then multiply that times all the robes worn by all the high priests throughout the history of Israel, we come to the number 365, one pomegranate for each day of the year.

These two significant numbers are just two suppositions. We will now look at the individual Biblical references to the pomegranate.

The first reference to the actual pomegranate tree, as opposed to the fruit, is found in I Samuel 14. Here we read, *"And Saul tarried in the uttermost part of Gibeah under a pomegranate tree which is in Migron: and the people that were with him were about six hundred men." (I Samuel 14:2)*

To get the history behind this verse, let's back up and look at chapter 13. We find here that God had just rejected Saul as king over Israel. Saul was told, *"But now thy kingdom shall not continue." (I Samuel 13:14)*

Saul's spiritual decline is also seen earlier in the chapter. Saul's son, Jonathan, won a victory over the Philistines. *"And Jonathan smote the garrison of the Philistines that was in Geba, and the Philistines heard of it." (I Samuel 13:3)* Though Jonathan won this victory, yet in the rest of this verse and the next one, we find that Saul took the credit for his son's victory. *"All Israel heard say that Saul had smitten a garrison of the Philistines." (I Samuel 13:4)*

It is this same thing that is happening again in chapter 14. Jonathan, with only one other man, his armor bearer, is where he should be, fighting battles. Even with this very small number, he believes *"there is no restraint to the LORD to save by many or by few." (I Samuel 14:6)* Yet while Jonathan is about his noble duty, all the while Saul is sitting under a pomegranate tree in Migron. He is doing nothing, yet he is taking the credit for all that his son does.

Interestingly not only is Saul sitting *"under a pomegranate tree,"* but he is also *"in the uttermost part of Gibeah."* Saul has insulated himself from the action. He is apparently hiding from the Philistines. Perhaps he originally stopped there originally for the pomegranate fruit, but it has now become a place to hide. The name "Migron" signifies a rocky precipice. This rocky territory is an ideal place to take shelter and hide. With the fruit of the pomegranate tree there for food, this makes a perfect place to camp while in hiding.

Migron's climate is ideal for having ripe pomegranate fruit at the time of year that Saul is there. Pomegranates grow best in subtropical climates, places with hot, dry summers and only cool winters. Indeed they can produce fruit and ripen it only in summers with high temperatures. If the temperature doesn't get high enough, the fruit will not ripen. Some small pomegranates can sometimes be grown in cooler temperatures, but not the large, lush pomegranates. Migron, a place with hot, dry summers, is the ideal place for growing these large pomegranates, and so at the time Saul hides there, its pomegranate fruit is large, lush, and ripe.

Saul's hiding amidst this luxury reveals his cowardice in contrast to Jonathan's courage. Jonathan, the one whose example we should follow, is not hiding. He has a firm trust in God's protection. He knows that the Lord will deliver him from the Philistines, and so he courageously takes their garrison. But Saul cowardly hides from the danger and then takes the credit for the victory afterwards.

The book of Joel also mentions the pomegranate tree when Joel describes the judgment upon Israel. *"The vine is dried up, and the fig tree languisheth; the pomegranate tree, the palm tree also, and the apple tree, even all the trees of the field, are withered: because joy is withered away from the sons of men."* *(Joel 1:12)*

All the fruits in this verse are sweet fruits. Often throughout the Bible the pomegranate fruit is mentioned along with these other sweet fruit trees, especially the vine, the fig, and the palm.

The fact that these sweet trees are withered is a mark of divine judgment. When there is no fruit on the trees, this means that they will not be able to produce fruit for many years to come. But mercifully, divine judgment is not necessarily permanent. Even these withered trees will produce fruit once again, though it will take several years.

The same is true with God's judgment upon His Own people. It is not necessarily permanent. After He finishes the judgment and removes it, we can produce fruit again. But we don't produce fruit right away. It takes several years for us to return to where we were.

The insects God uses to bring judgment in Joel are an interesting part of his prophecy. They are a picture of the invading armies that God uses to bring judgment to Israel, and are listed in verse 4 of chapter 1. *"That which the palmerworm hath left hath the locust eaten; and that which the locust hath left hath the cankerworm eaten; and that which the cankerworm hath left hath the caterpiller eaten."* *(Joel 1:4)*

These are probably not native insects, but foreign insects and invading species. One clue to this is in verse seven, which says that they eat the bark of the fig tree. This bark has a lot of acid, which insects don't like. Normally insects stay away from the bark of the fig tree. But for this judgment, God seems to raise up foreign insects that eat the acid-filled fig bark anyway. This seems to indicate that it is a foreign enemy that comes in and conquers Israel.

The judgment is thorough. It touches every tree, as Joel records. *"The flame hath burned all the trees of the field." (Joel 1:19)* But though these trees are destroyed for the time being, they are not totally eliminated. Their roots are left in the ground, and they can come back.

Chapter 2 of Joel offers hope through repentance, saying, *"Turn unto the LORD your God: for he is gracious and merciful." (Joel 2:13)* When the people heed this call and repent, the trees spring up again from their roots, and their fruit is restored. Joel records the joyful restoration. *"Be not afraid, ye beasts of the field: for the pastures of the wilderness do spring, for the tree beareth her fruit, the fig tree and the vine do yield their strength." (Joel 2:22)*

Just as He did here for Israel, God sometimes has to judge us for our sins. But if we, like these people, repent of our sins, then though it may take several years, God will restore our fruit again.

The first place pomegranate fruit is mentioned in the Bible is in Exodus 28:34 for the famous use of lining the high priest's robe. God commanded that the high priest's robe have *"a golden bell and a pomegranate, a golden bell and a pomegranate, upon the hem of the robe round about." (Exodus 28:34)* Almost the exact same wording is used in Exodus 39:26 where this command was obeyed. Here the children of Israel placed *"a bell and a pomegranate, a bell and a pomegranate, round about the hem of the robe to minister in; as the LORD commanded Moses." (Exodus 39:26)*

These bells were made of pure gold. As such, they symbolized the purity of the Gospel and of God's Word.

They were also a call to worship. When the high priest put on this robe and walked, these bells rang out, letting all Israel

Christ taught this truth clearly with this saying, *"But seek ye first the kingdom of God, and his righteousness; and all these things shall be added unto you." (Matthew 6:33)* Only if we seek Him and His righteousness first, the buds, will *"all these things,"* the fruit, be added unto us.

The final reference to the pomegranate that we will look at in the Song of Solomon is in chapter 8. Here the bride says to her beloved, *"I would lead thee, and bring thee into my mother's house, who would instruct me: I would cause thee to drink of spiced wine of the juice of my pomegranate." (Song of Solomon 8:2)*

By offering spiced pomegranate wine, this bride is giving the best she has to her beloved. With the pomegranate having 613 seeds, it takes a lot of time and effort to extract any quantity of juice from it. But besides this, she also takes the extra time and puts forth the extra effort to spice the wine. This is the very best she has to offer.

So we, like this bride, should give our best to Christ. We are pictured as His bride in the Bible, and He, the perfect Bridegroom Who gave His all for us, deserves our very best.

We will conclude by looking at the last mention of the pomegranate in the Bible, in the book of Haggai. The Lord said to the children of Israel, through His prophet Haggai, *"Is the seed yet in the barn? yea, as yet the vine, and the fig tree, and the pomegranate, and the olive tree, hath not brought forth: from this day will I bless you." (Haggai 2:19)*

Just as in other passages, we find that the pomegranate was listed with other sweet fruit trees, the vine and the fig. But it was also listed with another fruit tree that was not sweet, the olive tree. This verse, therefore, encompassed not only the sweet fruits, but the entire harvest.

To understand exactly what the Lord was saying here, let us look earlier in the chapter. The people had been unclean, as God demonstrated with this illustration. He told Haggai, *"Ask now the priests concerning the law, saying, If one bear holy flesh in the skirt of his garment, and with his skirt do touch bread, or pottage, or wine, or oil, or any meat, shall it be holy?"* Haggai asked them this question, and then *"the priests answered and said, No."* (Haggai 2:11-12)

These meats taught an important principle. That which is holy doesn't make holy that which is unclean. Rather that which is unclean defiles that which is holy. Just like these meats, the children of Israel had been unclean. Thus, even when they had offered "holy" sacrifices, their uncleanness had defiled their offerings. So they and everything they had touched had been unclean.

Because of this, God had sent leanness to their land. God described His judgment with these words, *"When one came to an heap of twenty measures, there were but ten: when one came to the pressfat for to draw out fifty vessels out of the press, there were but twenty. I smote you with blasting and with mildew and with hail in all the labours of your hands; yet ye turned not to me, saith the LORD."* (Haggai 2:16-17)

In some ways, these people compounded their judgment by having high expectations. When we expect to receive much and don't get it, then so much greater is our disappointment.

Yet this leanness was exactly what these unclean Israelites deserved. They had not fully cared for all of God's interests. Though they had performed their religious rituals, they had done so with unclean hearts. Therefore, God had received less than He deserved. So God, in return, did not take care of all their interests either. He gave them less than they expected.

This is the way God deals with us. When we don't do all that He asks of us, then He doesn't do all that we desire of Him either. When we serve Him only partway, He blesses us only partially as well.

But in Haggai 2, for the children of Israel, the time of judgment was now past. God said in verse 19, *"Is the seed yet in the barn? yea, as yet the vine, and the fig tree, and the pomegranate, and the olive tree, hath not brought forth: from this day will I bless you."* (Haggai 2:19)

The answer to God's question was, *"Yea."* Yes, their seed was yet in the barn, seed of all kinds, both sweet fruits like the pomegranate and non-sweet like the olive. And so the Lord started to bless them from that very day, and to allow them to recover what they had lost.

We truly serve a gracious God. Though He withholds blessings during times of judgment, He does so only to bring us to repentance. All the while His desire is to give us His fullest blessings. The moment we repent, He is quick to reverse His judgment and not only to bless us afresh, but even to restore what we have lost during the judgment. Of this sweet God, Who shed His blood in order to give us life, the sweet and red pomegranate is a wonderful picture.

38
The Poplar Tree

The poplar tree appears only twice in the Bible. The first time is in the account of Jacob working with Laban's flocks and herds. This passage says, *"And Jacob took him rods of green poplar, and of the hazel and chesnut tree; and pilled white strakes in them, and made the white appear which was in the rods." (Genesis 30:37)*

This poplar is called *"green poplar,"* but this is not the species green poplar. It refers rather to the fact that the tree is young and full of sap, like we use the term green wood. Its species is really white poplar.

Interestingly even the species that we know as green poplar, also known as yellow poplar, is not really a poplar at all. It is in the Latin classification *Liriodendron*. It is called both a green poplar and a yellow poplar. But the Biblical poplar is not this green poplar. It is rather a true poplar under the Latin classification *Populus*. Its Hebrew name is *libneh* (לבנה).

We will now look at the only other Biblical use of the poplar. Hosea says, *"They sacrifice upon the tops of the mountains, and burn incense upon the hills, under oaks and poplars and elms, because the shadow thereof is good: therefore your daughters shall commit whoredom, and your spouses shall commit adultery." (Hosea 4:13)*

We have looked at this verse in the chapters on the oak and the elm, but now we look at in in relation to the poplar. We find the poplar here, like the oak and the elm, connected with heathen idol worship. Even beyond this verse, this connection with heathen worship can be seen in secular history. The ancient Greeks had a legend that Hercules, one of their gods, had originally brought the poplar tree to Greece.

Some people argue that the poplar tree cannot grow on mountains and therefore cannot be the tree mentioned in this verse. They say that this tree is not truly a poplar, but a storax. But let's look further.

First, the storax is a very small tree or shrub, not tall enough to burn incense under. Second, though the storax does grow in mountainous regions and dry, rocky places, this is not a requirement for the poplar tree in this verse. The verse says that the people sacrifice on the mountains, but not that they burned incense on the mountains. They burn incense *"upon the hills."*

In this light, I believe the poplar tree referred to is the true poplar, probably the white poplar. The white poplar is the one that is the most drought tolerant of all poplars, and therefore can grow best on hillsides. It can be seen today on forest edges where there is less water than most poplars require for growth.

So we find that the poplar tree in the Bible is not necessarily mentioned in a good light. Even in the passage on Jacob's rods, Jacob was using the poplar deviously in order to trick Laban and increase his own flocks and herds above those of Laban. So the poplar, like the oak and elm, reminds us that nothing is inherently good in itself. Its worth is determined by its use. Let us, therefore, be careful how we invest our own goods and our own lives, so that they might be truly good and useful in the service of the Lord.

39
Pruning

In John 15, Christ taught on the process of pruning as it applies to the human life. Perhaps surprisingly, the word "prune" is not found here, nor is it found in any New Testament passage. The word "prune" is found only in the Old Testament, but the concept is expounded here by Christ Himself.

Christ began by saying, *"I am the true vine, and my Father is the husbandman." (John 15:1)* The context of this discourse is the Last Supper. Christ was about to leave the world. He was concerned that, when He did, His disciples not depart from what He had taught them. For this reason, He told them that He was the true vine. He did not want them to go back to a false vine, such as the law of Moses.

By likening Himself to the vine, Christ was indicating something of the need to spread the Gospel. The vine is a spreading plant. Its branches spread out and reach far from where they began. In the same way, Christ was soon to give the Great Commission, commanding Christians, the branches, to spread the Gospel of Jesus Christ throughout the whole world. Since that day, just as the vine spreads, Christ's name has spread all throughout the world.

Christ then pictures God the Father as the husbandman. The husbandman of any field must take care of his field, pruning, mowing, growing the crops, striving toward production of fruit. God is the perfect husbandman, the perfect vinedresser, and His vine is Christ, the perfect vine.

Throughout this passage, there is a progression in the amount of fruit borne by the branches. Verse 2 says, *"Every branch in me that beareth not fruit he taketh away: and every branch that beareth __fruit__, he purgeth it, that it may bring forth __more fruit__." (John 15:2)* As young, immature Christians, we bear some fruit. As we grow in Him and are purged, we bear more fruit. Finally in verse 5, the mature Christian bears much fruit. *"I am the vine, ye are the branches: He that abideth in me, and I in him, the same*

254

*bringeth forth **much fruit**: for without me ye can do nothing."* *(John 15:5)*

Christ wants us not only to bear some fruit, but more fruit, and eventually much fruit. The more Christlike we become, the more fruit we will bear. Throughout the Christian life, we should continue to bear more and more fruit, so that at the end of life we are bearing as much as we can.

If we do not bear fruit as we should, Christ warns that we will be pruned, purged until we bring forth more fruit. *"Every branch in me that beareth not fruit he taketh away: and every branch that beareth fruit, he purgeth it, that it may bring forth more fruit." (John 15:2)*

To purge means to cleanse. The only way to cleanse the vine of an unfruitful branch is to remove the branch. Purging is the act of removing. Cleanliness is the result that comes from that act. Thus where Christ says *"he purgeth it,"* it means that He removes it.

In grafting there were two ways to remove a branch, breaking and cutting. Here are two more, purging and pruning. While all of these are similar, each is slightly different from the others. Breaking is the Greek word *klao* (κλαω), which simply means to break. It is used of breaking bread, not a careful, clean cut, but a rough, jagged break. Cutting is the Greek word *kopto* (κοπτω) and means to chop or strike. It carries with it the picture of a clean, smooth cut. Purging, the process found here in John 15, is the Greek word *kathairo* (καθαιρω) and means to cleanse. From this word, we get the name Katherine or Kathy. It focuses on cleansing, removing all evidence of the bad branch. Pruning, found only in the Old Testament, is the Hebrew word *zamar* (זמר) and means to trim. Within the pronunciation of this word (zaw-mar') can be heard the English word "saw." It is the trimming, the sawing off, of the branch.

The branches spoken of here of course are people, not literal tree branches. I Corinthians says, *"For we are labourers together with God: ye are God's husbandry, ye are God's building." (I Corinthians 3:9)* God's husbandry is not plants, but living people. We make up His vineyard.

Though it is contrary to natural thinking, cutting off a part of a plant actually makes it more fruitful. A petunia plant is grown for its flowers. But the way to cause many flowers to grow on the petunia is by pinching off the first flowers. In the same way, the Lord must pinch off some things in our lives, in order that we may bring forth more fruit.

Hebrews 1:3 speaks of the Lord's purging of our sins. *"Who being the brightness of his glory, and the express image of his person, and upholding all things by the word of his power, when he had by himself **purged** our sins, sat down on the right hand of the Majesty on high." (Hebrews 1:3)* Christ removed our sins, got rid of the sin that was in us.

Even after salvation, the branches must continue to be purged. For the branches are a part of Christ, the true vine, and nothing un-Godly or unholy can be a part of Christ. Therefore sin must continually be purged from the body of Christ.

Malachi 3 contains the only mention of purging in the Old Testament. It says, *"And he shall sit as a refiner and purifier of silver: and he shall purify the sons of Levi, and **purge** them as gold and silver, that they may offer unto the LORD an offering in righteousness." (Malachi 3:3)* This verse is speaking of the removing of impurities from metal. The more impurities that are removed, the stronger the metal. In the same way, impurities make weak Christians. The Lord must purge out these impurities and cleanse us in order to make us stronger. To translate the analogy to trees, if a branch isn't purged of its impurities, it will affect the whole tree. As branches, we must be purged, purified, in order that we not weaken and corrupt the rest of the tree.

The cleansing that we have in Christ is spoken of throughout the Bible. Ephesians says, *"That he might sanctify and cleanse it with the washing of water by the word." (Ephesians 5:26)* I John likewise says, *"If we confess our sins, he is faithful and just to forgive us our sins, and to cleanse us from all unrighteousness. (I John 1:9)* Though we are saved, we are still sinners. We must continually be cleansed, purged of our sins. This is the occupation of God, the vinedresser, to purge the unfruitful branches continually that they might bring forth more fruit.

Verses 3 and 4 of John 15 go on to speak of our abiding in Christ. Abiding is the attachment between the Christian and Christ, like the attachment of a branch to a vine. However, it must be remembered that this passage is talking about people, not real plants. The attachment between us and Christ is not a physical attachment, but a spiritual attachment. We are not physically a part of Christ. We are not gods.

The best illustration of this is the Biblical example of marriage. In Genesis 2:24, God says of a man and his wife, *"Therefore shall a man leave his father and his mother, and shall cleave unto his wife: and they shall be one flesh." (Genesis 2:24)* This does not mean that they become one person physically. When we look at them, we still see two people. They are joined together in heart and mind and spirit.

In the same way, when others see us, they should not see Jesus Christ physically. But they should see the nature of Jesus Christ in us. We should be joined to His character and spirit, as a branch is joined to a vine.

As with other analogies, we must be careful not to take this too far. A willow branch, for example, can be planted in the ground on its own, and it will grow. It does not need to be attached to the trunk in order to live. Thus not all tree-branch relationships are exactly like the relationship between Christ and the church. We must be careful to take the analogy as it is presented in the Bible and use it as it is intended there. Yet when used carefully, much can be learned from the characteristics of literal plants that can be applied to the Christian life.

For example, it is very instructive to look at the reasons for pruning real plants. These reasons reveal many reasons the Lord has for pruning the unfruitful branches from the vine of Christ. The ultimate reason is given in John 15:2, *"that it may bring forth more fruit." (John 15:2)* But there are several individual causes of unfruitfulness that are seen both in real branches and in Christians.

The first reason to prune branches is to shape the tree into proper form. This varies from tree to tree. An apple tree should have broad branches that hang low, and an open effect. A shade tree should be just the opposite. Its branches should not be broad, but high and thick. God prunes Christians according to the form of Christ.

Another time to prune is when two branches cross one another. As the wind blows, the two branches rub against one another until they rub the bark off. Once the bark is gone, insects and disease have access to the branches, and both are killed. In order to prevent such a tragedy, one branch must be removed.

As God looks upon the vine of Christ, He at times sees two crossing branches, two Christians rubbing against one another. This is because one of the branches is growing in the wrong direction, and therefore one must be removed. Otherwise both will die. This is the reason for church discipline, a difficult, but necessary, course of action.

The crossing branches may also be larger than individuals. They may be churches rubbing against one another. They may even be whole denominations rubbing against one another. In any case of crossing branches, it is God Who decides which branch to remove. He selects the perfect one, according to the form of Christ.

Dead branches are yet another cause for pruning. Dead branches allow insects and disease into the tree. Though physically attached to the tree, they are not truly a part of it. In the same way, dead spiritual branches claim to be abiding in Christ, but they are not partaking of His spiritual life. They neither give nor take sustenance from the tree, the body of Christ, yet their presence affects the whole, both corporately and individually.

Often on a tree it is the lower branches that are dead because they receive no sunlight. Jesus Christ is the sunlight of spiritual branches. This is seen in Revelation where physical sunlight is no longer needed, for He is the perfect sunlight. *"And there shall be no night there; and they need no candle, neither light of the sun; for the Lord God giveth them light: and they shall reign for ever and ever."* *(Revelation 22:5)* If we do not receive Jesus Christ as our sunlight, as the Light of the World, we will become dead branches as well. For His sunlight is vital to our life and growth.

An evidence of a dead branch is the lack of fruit and leaves. When a branch ceases to produce fruit, it soon ceases to have leaves. And when it has no more leaves, it is dead. It is the same with the Christian life. When a Christian ceases to produce fruit, he also ceases to obey the Lord in certain areas. He may begin with missing church services, until eventually he has lost all his leaves, his acts of obedience. Then he is spiritually dead, and Christ must remove him in His righteous judgment. Judgment is always necessary in determining which branch to remove, and only Christ has perfect judgment.

Another occasion when pruning is necessary is when two central branches are growing up side by side. Both compete for headship. If they are allowed to continue in this way, the tree will eventually split apart as if a wedge were driven into it. One of these must be removed to prevent the stress caused by such a division.

The Lord has established a certain authority structure. In a church there is to be one pastor. In a marriage there is to be one head, the husband. In the Christian body as a whole, we have only one Lord. When there is competition for that headship, there is confusion among those that are to be led. If one is not removed, there will eventually be division as one breaks off. Then it will not be the neat cut of pruning, but an ugly break that wounds the other branch and takes vital substance with it.

The time to prune a branch is when it is young. If there is a branch that is in some way harming the tree, not having proper form, rubbing against other branches, being unfruitful and dead, or competing for headship, this problem must not be allowed to fester. The longer the branch is allowed to grow, the harder it will be to prune it. The larger the branch, the larger the wound it leaves when it is removed. When one has remained in the church a long time and then is removed for sin, it damages all those to whom he has ever ministered in his history. It leaves a deep wound in the whole church. The way to prevent such wounding and to minimize the damage to the tree is to prune the branch when it is young.

The timing of removal is vital to proper pruning. For example, an tree should not be pruned in the late spring or summer. Otherwise the calloused end will not heal before the time of fruit, and the fruit production will be limited. Rather a tree should be pruned in the fall. Then the calloused end has all winter to heal, and by the next year, it will bear more fruit. In spiritual pruning, God the Father is the perfect husbandman. He knows precisely the right time to prune a branch. When a Christian falls into sin, he should quickly repent and not continue in it, so that he will not need to feel God's pruning. However if he does continue in it, God knows when he has gone beyond the point of repentance, and God knows the proper time to remove him.

Related to the pruning of branches is root pruning. Root pruning is described in Luke 13, in which Christ gives a parable of a fig tree, *"A certain man had a fig tree planted in his vineyard; and he came and sought fruit thereon, and found none. Then said he unto the dresser of his vineyard, Behold, these three years I come seeking fruit on this fig tree, and find none: cut it down; why cumbereth it the ground? And he answering said unto him, Lord, let it alone this year also, till I shall dig about it, and dung it. And if it bear fruit, well: and if not, then after that thou shalt cut it down."* (Luke 13:6-9)

When a vinedresser digs about a tree as in this parable, he cuts the roots. This retards the growth of the tree and is a great shock to the tree. But in severe cases this is exactly what it needs. For the very shock of this process can sometimes shock a tree into producing fruit. Of course, not all the roots can be pruned, or the tree will die. This is a severe measure, but is sometimes necessary in order to obtain fruit.

After Christ explains the need for pruning, He goes on in verse 4 of John 15 to show the necessity of abiding in Him. *"Abide in me, and I in you. As the branch cannot bear fruit of itself, except it abide in the vine; no more can ye, except ye abide in me."* (John 15:4) As Christians we must abide in Christ, for abiding is the vital attachment of the branch to the tree. Verse 6 warns of the tragic consequences of not abiding in Christ. *"If a man abide not in me, he is cast forth as a branch, and is withered; and men gather them, and cast them into the fire, and they are burned."* (John 15:6)

Who is this verse talking about? Some have interpreted these branches to be the lost who are burned in Hell. They are in the church, seemingly abiding in Christ, but are unsaved. Because they are dead branches, they bring forth no fruit. They have a form of Godliness, but are not saved. Therefore they are purged from the church and eventually burned in Hell fire.

This seems plausible, but the context of this passage must be kept in mind. This is during the Passover, when Christ is talking to His true disciples. Judas is not among them. After Jesus had given him the sop, chapter 13 says that he *"went immediately out: and it was night." (John 13:30)*

Verse 4 also says, *"Abide in me, and I in you." (John 15:4)* Those who abide in Christ simultaneously have Christ abiding in them. This is a clear picture of the saved soul, for if Christ dwells within a soul, that soul is saved. Verse 5 says the same thing, *"He that abideth in me, and I in him, the same bringeth forth much fruit." (John 15:5)* Again verse 7 says, *"If ye abide in me, and my words abide in you, ye shall ask what ye will, and it shall be done unto you." (John 15:7)* If we abide in Christ, Christ and His words abide in us. If Christ abides in us, we are saved.

The burning spoken of in this passage cannot be the burning of a soul in Hell. Once a soul is saved, once he abides in Christ and Christ abides in him, he cannot lose his salvation. The unsaved on the other hand are not indwelt by Christ. They do not have Christ abiding in them. They cannot be purged from the tree, for they never did abide in Christ and never were saved. Jesus will say to them, *"I never knew you." (Matthew 7:23)* He doesn't say, "I once knew you, and now I don't," but "I never knew you." These will suffer burning in Hell, but that is not what is spoken of here.

The branches that are purged from the vine of Christ in this passage are instead the works of Christians. This principle is taught in I Corinthians 3. *"Now if any man build upon this foundation gold, silver, precious stones, wood, hay, stubble; Every man's work shall be made manifest: for the day shall declare it, because it shall be revealed by fire; and the fire shall try every man's work of what sort it is. If any man's work abide which he hath built thereupon, he shall receive a reward. If any man's work shall be burned, he shall suffer loss: but he himself shall be saved; yet so as by fire. Know ye not that ye are the temple of God, and that the Spirit of God dwelleth in you?" (I Corinthians 3:12-16)*

This passage in I Corinthians is written to Christians. It is for those who have the Spirit of God dwelling in them, just as John 15 is for those who have Christ abiding in them. Here the Christian's evil works will be burned, but *"he himself shall be saved." (I Corinthians 3:15)* The burning of the branches cannot refer to eternal damnation, but rather to the burning of our works.

It must be said that some Christians do die prematurely when they continually do not produce fruit. But their souls are not cast into Hell. They themselves are saved, *"yet so as by fire." (I Corinthians 3:15)* Their works are burned, but their souls are saved from the fire.

These passages in John 15 and I Corinthians 3 warn us as Christians to be very careful in our works. Often what we think we're doing for Christ, we're really not doing because we have a corrupt motive. It all comes down to what our motives are. The danger of which we are warned is being attached to Christ by salvation, but not abiding in Him, drawing upon His nutrients, in daily living.

Matthew 7 uses the burning of branches in a different analogy to illustrate the burning of lost souls in Hell. *"Ye shall know them by their fruits. Do men gather grapes of thorns, or figs of thistles? Even so every good tree bringeth forth good fruit; but a corrupt tree bringeth forth evil fruit. A good tree cannot bring forth evil fruit, neither can a corrupt tree bring forth good fruit. Every tree that bringeth not forth good fruit is hewn down, and cast into the fire. Wherefore by their fruits ye shall know them." (Matthew 7:16-20)*

Matthew says nothing about abiding in Christ. In fact the good tree and bad tree in this passage are totally different trees, one likened to a saved soul, the other to an unsaved soul. These are not branches abiding in the vine, but entire trees themselves. So there is nothing to prune. There is only the removal of evil fruit and the saving of good fruit.

The evil tree produces evil fruit, just as evil motives in the unsaved heart produce evil fruit. This reveals a great truth. Evil fruit is produced because the tree is evil, not the other way around. Men sin because they are sinners. They are not sinners because they sin.

Christ will reject this evil fruit of the unsaved heart. He will cut down the tree and cast it into the fire. The fire in this passage is Hell fire. The evil fruit demonstrates that this tree never was a part of the good tree, Christ.

This evil fruit can be illustrated in the world of nature by the honeysuckle plant. Its bright red berries in the fall and winter look beautiful and good to eat, but they are poisonous. While they are not poisonous enough to kill a person, they can make him so sick he wishes he were dead. Such destructive poison is the evil fruit of the unsaved soul. It is this evil fruit that reveals the poisonous, evil nature of the soul, which must be cast into Hell fire.

This interpretation of Matthew is not only consistent with the passage itself, but also with its entire context. Verses 13 and 14, the verses just before the passage, say, *"Enter ye in at the strait gate: for wide is the gate, and broad is the way, that leadeth to destruction, and many there be which go in thereat: Because strait is the gate, and narrow is the way, which leadeth unto life, and few there be that find it."* (Matthew 7:13-14) These verses clearly refer to the saved and unsaved. The strait gate leads to Heaven, but the broad gate to Hell.

Also the verses just after the passage on trees continue in this same vein. *"Not every one that saith unto me, Lord, Lord, shall enter into the kingdom of heaven; but he that doeth the will of my Father which is in heaven. Many will say to me in that day, Lord, Lord, have we not prophesied in thy name? and in thy name have cast out devils? and in thy name done many wonderful works? And then will I profess unto them, I never knew you: depart from me, ye that work iniquity."* (Matthew 7:21-23) Again the context here is the difference between the saved and unsaved.

Since the verses both before and after the passage on trees are talking about salvation and the eternal destruction of the lost in Hell, we have to conclude that the verses in between are talking about that also. This keeps the passage in context and keeps us true to the meaning of Scripture.

To finish our study of pruning, we take a final look at the few Biblical mentions of the actual word "prune." While some of the newer translations translate the word "purge" in John 15 as "prune," this fails to capture the idea of cleansing. The real word for pruning, "zamar," appears only a very few times in the Bible. It was surprising to find how few actual occurrences of this word there are.

Leviticus mentions pruning when it institutes the law for giving the land a Sabbath rest. *"Six years thou shalt sow thy field, and six years thou shalt prune thy vineyard, and gather in the fruit thereof; But in the seventh year shall be a sabbath of rest unto the land, a sabbath for the LORD: thou shalt neither sow thy field, nor prune thy vineyard." (Leviticus 25:3-4)* During the seventh year, pruning, along with all other husbandry activity, was forbidden.

Isaiah mentions pruning in describing God's judgment upon Judah. Likened to a vineyard that has failed to produce fruit, Judah must bear the judgment of God ceasing to prune and care for the vineyard. *"And I will lay it waste: it shall not be pruned, nor digged; but there shall come up briers and thorns: I will also command the clouds that they rain no rain upon it." (Isaiah 5:6)*

The only other references to pruning refer to pruninghooks. Isaiah 18 describes the use of the pruninghook. *"For afore the harvest, when the bud is perfect, and the sour grape is ripening in the flower, he shall both cut off the sprigs with pruning hooks, and take away and cut down the branches." (Isaiah 18:5)*

The pruninghook is in reality a knife. The difference is that it is on a long pole and is bent over into a hook on the end. This makes pruning easier because the knife goes over the limb a little in order to saw it off. If one goes to the store today to buy a pruninghook, he will find it to be like the Biblical pruninghook, a hooked knife on a pole. A roofing knife has a similar shape.

Joel uses the pruninghook to describe the coming days of great war during the Tribulation, when the nations will make war against Israel. *"Beat your plowshares into swords, and your pruninghooks into spears: let the weak say, I am strong." (Joel 3:10)* The spear, a strait knife on a long pole, can be easily beaten from a pruninghook.

This law of God represents to us today the entire Word of God. Just as the ark showed the direction in which the children of Israel were to go, so the Bible shows us today the direction in which we are to go. Wherever the Bible goes, we are to go. We should hide God's Word in the arks of our hearts, and follow wherever it tells us to go. The wood overlaid with gold is not only a picture of Christ, but also of us as Christians. We are earthly and human, like wood from a dry ground. But when we trusted Christ as Savior, we were made incorruptible in Him. We were not made perfect in our daily lives, but we were justified in the sight of God through the perfection of Christ. And we were overlaid with pure gold. This is not our incorruptibility or our purity, but Christ's. Thus through faith in Christ, we receive His incorruptibility and His purity, and we also receive all the additional riches and blessings that are contained in Him, as pictured by the other items in the ark, Aaron's rod that budded and the pot of manna.

On top of the ark was a lid. This lid, however, was not made of shittim wood overlaid with gold, but was made of solid gold. This beautifully illustrates the fact that the pureness of Christ covers all of our sin. His pure blood, untainted by any sin of its own, covers all. No matter what is beneath, solid and pure gold is on top and is all that the holy God above sees.

The ark, the picture of God Himself, was the only piece of furniture in the most holy place, or the holy of holies. It was the most important piece. Wherever it went, the entire tabernacle and entire congregation of Israel followed.

In verse 13 of Exodus 25, we read of the staves that were made to transport the ark. *"And thou shalt make staves of shittim wood, and overlay them with gold." (Exodus 25:13)* These staves were rods that went through rings attached to the ark, and were used to transport it.

The priests could touch these staves only, not the ark. In the time of David, Uzzah touched the ark, not maliciously, but impulsively, trying to steady it. For even this, he was struck dead. This severe punishment for even an "accidental" touching reveals that this was a very serious matter. Man was not to touch that which was holy.

These staves represent us as Christians. Like the ark pictured, we are wood, or human clay, but we are overlaid with gold, the purity and salvation of Jesus Christ.

While the ark gave the direction of travel, the staves still had to carry it. In the same way, Christ gives us the direction in which we are to go, but He is not going to do the work for us. We still have the responsibility to walk in the way He shows us and to carry Him and His Word wherever we go.

Of course, in another sense, Christ carries us. With analogies, there is always a danger of carrying them too far. For this one, we must not carry it too far and say that we sustain Christ. In reality, He sustains and carries us. We must also realize that the ark was not truly Christ, but only a type of Him. So let us use this analogy properly, and allow it simply to reveal to us our responsibility to carry the bright testimony of Christ everywhere we go.

These staves were not to be taken out of the ark, as we read further. *"The staves shall be in the rings of the ark: they shall not be taken from it." (Exodus 25:15)*

This was so that, at a moment's notice, when God gave the order to move, the children of Israel would be able to pick up the ark and go. They wouldn't have to go hunt for the staves first. In the same way, we should be ready at a moment's notice. When Christ leads us anywhere, perhaps to go witness to someone at a gas station or grocery store, we should not put it off. We should, like these staves, already be ready to go and should go immediately.

The next piece of tabernacle furniture for which shittim wood was used was the table of shewbread. God commanded, *"Thou shalt also make a table of shittim wood: two cubits shall be the length thereof, and a cubit the breadth thereof, and a cubit and a half the height thereof." (Exodus 25:23)*

This was not a tall table, but was just over two feet tall. Like the ark, it was overlaid with gold, picturing Christ's purity coupled with His humanity. It also, like the ark, pictured man's human nature covered by the purity of Christ.

The twelve loaves of bread on the table may picture the twelve tribes, or maybe the twelve apostles, or maybe both. There altar was likewise lifted up, and there it consumed many sacrifices. But the difference between Jesus Christ and this altar is that, while the altar consumed many sacrifices, Jesus Christ died only once as a sacrifice.

This altar was outside the holy place of the tabernacle, and instead was in the open air courtyard. Because it was used to burn the animal sacrifices, it was necessary that it be outside so the smoke and fire would have a place to go. Being outside the tabernacle, the altar pictured the fact that Jesus Christ was crucified outside the temple precincts.

Also being outside, it pictured the fact that the death of Christ is not what we should continually carry in our hearts. We, like the tabernacle, are today the temple of the Holy Spirit, the very dwelling place of God. We should carry with us continually the things that were inside the tabernacle. We should have the ark, picturing Christ and His Word. And we should have the table of shewbread, picturing communion. But we shouldn't continually live with the death of Christ in us. We should live rather with His life in us. He died once, but He lives forever. It is this eternal life that we must carry in our hearts.

This altar was not overlaid with gold, but instead with brass. I believe this is because this altar represents the death of Christ. Brass is not pure like gold, and thus pictures the fact that our impurities and sins were laid upon Christ. Brass is also not long lasting like gold, picturing the fact that Christ died once, and did not remain on the cross. His death is not a continual sacrifice, but a once-for-all atonement. He is not still on the cross today, but is alive.

Brass is an alloy of copper and zinc, and thus I believe it represents two things, both judgment and atonement. On the cross, pictured by the brazen altar, both of these were accomplished. God poured out His judgment, not upon us as guilty sinners, but upon His Own Son. Because of this judgment, we receive the atonement of our sins through the same cross.

Being the highest piece of furniture, this altar was elevated to a place of prominence. Being outside, it was continually before the eyes of the people. It pictures the fact that Christ should be elevated and exalted in our hearts and lives.

On this altar, the fire consumed the sacrifices. But on the cross of Christ, the sacrifice consumed the fire. After the death of Christ, there was no more any need for the fire of animal sacrifice. Christ has now offered the perfect sacrifice, once for all. Though we should be living sacrifices and should die daily, yet we are not literal and physical sacrifices. These sacrifices have been forever ended with the perfect sacrifice of Christ.

The brazen altar stood just outside the curtain that covered the entrance to the holy place of the tabernacle. This was the only entrance. There was no side entrance or back door. Thus anyone who entered the sanctuary had to pass by this altar. In the same way, anyone who desires to enter heaven must pass through the cross of Christ. He is the only way, as He said, *"I am the way, the truth, and the life: no man cometh unto the Father, but by me."* *(John 14:6)*

Each piece of tabernacle furniture was placed in its position for a specific reason. We do the same thing with our furniture. When we move into a new house, we don't just throw our couch in the front door and leave it wherever it lands. No, we carefully place it in a specific place for a specific reason. God did the same thing with His furniture, and the position of each piece is important and instructive to us.

The brazen altar also had a horn on each of its four corners. God said, *"And thou shalt make the horns of it upon the four corners thereof: his horns shall be of the same: and thou shalt overlay it with brass."* *(Exodus 27:2)* *"Of the same"* means that the horns were made of the same wood as the rest of the altar, shittim wood. Upon these horns the meat of each sacrifice was hung and burned.

Horns are a symbol of strength. In Israel, the ram with his horns was a common sight and was a constant picture of this strength, as he strode majestically through the rolling pastures. We in the United States are more familiar with the deer, which pictures this same thing. A male deer with large antlers is a great symbol of strength and might.

Horns are also a symbol of exaltation. They are the highest feature on the animal and are thus exalted and lifted up. The horns of the brazen altar were the highest objects on the tallest piece of furniture. They were thus a very noticeable picture of exaltation.

The horns combined their symbolism with the altar to give us a beautiful picture of Jesus Christ. He Who is our strength and Who is exalted high above us and above all the earth, is the perfect sacrifice for us. He was exalted and lifted up on the cross and is the perfect and complete atonement for our sins.

The brazen altar, like the ark and the table of shewbread, also had staves by which it was carried. *"And thou shalt make staves for the altar, staves of shittim wood, and overlay them with brass." (Exodus 27:6)*

These staves remind us that we as Christians are to carry the cross of Christ with us. We should not necessarily carry it physically around our necks, on our fingers, or in our vehicles. This can become Pharisaical and just a show. What is on the outside is not necessarily what is on the inside.

We should instead carry the cross on the inside, in our hearts, knowing that Christ died for us and living in the light of that truth. As we do this, we will carry the cross as a testimony to others of what the cross can do for them.

The staves also pictured Christ. They were shittim wood, humanity, overlaid with brass, both atonement and judgment. Just as the brass covered the wood, so Christ's atoning blood covers our sin, which was fully judged on the cross. Also just as the staves carried the altar, so Christ must carry us to heaven, for we can't get there on our own. But He carries only those who come through the cross, the altar of His sacrifice. This is our only way to heaven.

The last piece of furniture that was made of shittim wood was the altar of incense, mentioned in chapter 30 of Exodus. This chapter is worthy of our special attention because it is the only one in the Bible, aside from Matthew 2, that mentions all three of the gifts brought by the wise men to the child Jesus, gold, frankincense, and myrrh. This shows just how important the tabernacle is. It is one of the chief Old Testament representations of Christ, and understanding it is very important to our lives.

This chapter describes the altar of incense with these words, *"And thou shalt make an altar to burn incense upon: of shittim wood shalt thou make it."* *(Exodus 30:1)* The incense offered on this altar represents, not the once-for-all blood sacrifice of Christ, but our daily offerings of prayer and praise. Unlike the brazen altar, this altar was actually inside the holy place, not in the outdoor courtyard. Thus these offerings should be continually in our hearts.

But the prayers that we offer cannot go directly to God the Father. He is holy, and we are sinful. Therefore, we need an intercessor, someone to take our prayers from our lips and carry them to the Father, just as this altar did with the incense. This office can be fulfilled by only one, Jesus Christ, for He is both God and man. Thus this altar is yet another type and picture of Jesus Christ and reveals yet another of His many roles.

The altar, like the ark of the covenant, was made of shittim wood overlaid with gold. The incorruptible shittim wood pictured Christ's incorruptible and sinless humanity. The gold pictured His divine purity. By God's design, the shittim wood of His humanity was entirely covered, so that only the gold, the picture of divine purity, could be seen.

The altar of incense was placed just outside the curtain that led into the most holy place, the place that contained the mercy seat, the dwelling place of God Himself. No one standing at this altar could see the mercy seat. It was entirely enclosed by curtains. To reach it, one must first pass through the altar of incense.

This indicates that we cannot approach God the Father directly with our prayers. Though we can *"come boldly unto the throne of grace,"* *(Hebrews 4:16)* we cannot come on our own merits. We can come there only through Jesus Christ our intercessor.

Romans tells us that our prayers are unintelligible to God. It says, *"we know not what we should pray for as we ought: but the Spirit itself maketh intercession for us with groanings which cannot be uttered."* *(Romans 8:26)* Just as we cannot understand or utter these groanings of the Spirit, so our prayers sound like unintelligible groanings and mutterings to God. It is for this reason that Christ makes intercession for us to God the Father.

42

The Sycamore Tree

The sycamore tree, like the sycamine, is in the fig family. The members of this family can be ranked in order of importance, with the common fig being the most important, the sycamore the next, and the sycamine the least.

The Greek word for the sycamore is *sukomoraia* (συκομοραια), a different word than that translated sycamine, *sukaminos* (συκαμινος). The Biblical sycamore tree is not like our North American sycamores. Our sycamores are really planetrees, with the Latin name *Platanus*, which means "plane." Our trees are tall with whitish, flaky bark and large leaves.

But the Biblical sycamore is much like both the fig and the mulberry. In fact its Greek name *sukomoraia* (συκομοραια) comes from the word *sukon* (συκον) which means "fig," and the word *moron* (μορον) which means "mulberry." It is sometimes called the fig mulberry. It gets the name "fig" because it is in the fig family, classified as *Ficus sycomorus*. And it gets the name "mulberry" because it looks like a mulberry. But it is not a true mulberry, which is in the classification *Morus*.

The sycamine has also been likened to both the fig and the mulberry. So to distinguish between the sycamine and the sycamore, the sycamine is more like the mulberry, and the sycamore is more like the fig. The sycamine resembles the mulberry in both its leaves and fruit, which fruit is small and dark like mulberry fruit. But the sycamore, though its leaves do resemble mulberry leaves, is more like the fig in its fruit. Its fruit is larger than the fruit of either the sycamine or the mulberry, and is sweeter. It is also yellowish gold.

But do not be confused. All four of these are different trees. They are called by different names in the Bible, and, while they have some features that are similar, they each have their own individual characteristics which set them apart from the others.

An interesting fact about the sycamore is that its fruit does not ripen unless it is first scraped or pierced. This can be done with an iron claw, with a hand rake such as would be used in a garden, or even with a simple knife. This piercing of the sycamore fruit causes it to emit ethylene gas, which quickens the ripening of any fruit. This gas is used commercially today to ripen fruit in stores, or it is withheld in order to retard its ripening until it reaches the shelf.

This brings us to an interesting mention of the sycamore is in the book of Amos. Because this is the Old Testament, the word "sycamore" comes from the Hebrew rather than the Greek. The Hebrew word for the sycamore is *shiqmah* (שִׁקְמָה).

In this book, Amos *"said to Amaziah, I was no prophet, neither was I a prophet's son; but I was an herdman, and a gatherer of sycomore fruit."* (Amos 7:14) The word "gatherer" refers to this very process of piercing the sycamore fruit. In fact, the Septuagint translates this phrase as "piercer of sycamore fruit." It really encompasses the entire activity of working with sycamore fruit, both piercing the fruit to hasten its ripening, and then gathering the ripe fruit later.

This is clearly the work of a poor and common man, which Amos was. Only the poor people ate this lesser quality sycamore fruit, while the rich people ate the sweeter figs. Amos was only a simple herdsman, and so he ate the common man's fruit. This is also why he did both the piercing and gathering. He didn't have servants to do these tasks for him. Instead, he did it all.

This teaches us that God uses the common and weak things to accomplish His will. Amos, just a common man, was talking to the powerful king of Israel, Amaziah. He was prophesying the destruction of both the king and his kingdom. Indeed God uses the lowly to bring down the great.

This truth is illustrated in the life of Christ. He came to earth in a humble, poor manger, yet He is Lord of all. He also did the work of the common man, the piercing and gathering of sycamore fruit. He allowed Himself to be pierced on the cross so that He might one day gather us, His fruits, home in heaven.

This same principle of God using the lowly is seen throughout the Biblical occurrences of the sycamore. The sycamore tree is often mentioned with other trees, always trees that are more important than itself. Here in Amos, more important and useful trees appear earlier in the book. Chapter 2 likens the Amorite to the powerful cedar, and chapter 4 likens Israel to the sweet fig. But both of these are destroyed through the words of a simple gatherer of the common sycamore in chapter 7. This is the lesson of the sycamore tree, that the lowliest being, when given God's power, is greater than even the mightiest force on earth that is without God's power.

Another mention of the sycamore that illustrates this truth is in I Kings 10:27. Almost the same wording is found in II Chronicles 1:15 and 9:27, but since they are so similar, we will look only at I Kings.

Here in I Kings, when Solomon gathered materials for the temple, he imported so many cedar trees from Hiram king of Tyre that the Bible says, *"Cedars made he to be as the sycamore trees that are in the vale, for abundance." (I Kings 10:27)* This reveals the fact that sycamores were very abundant in Israel in Biblical times. In other words, the sycamore was just a common tree that could be found anywhere, whereas the cedar was rare and valuable.

Thus even though the quality of sycamore fruit was poor, the quality of its wood was poorer. So this tree was used more for its fruit than for its wood. We have seen this with other trees. For example, the fig is useful for fruit, but its wood rots very easily. The shittah, or acacia, tree is the opposite. It has very useful wood, but not fruit.

For building the temple, the low quality sycamore wood was not suitable. So Solomon imported trees with higher quality wood, cedar trees and almug trees. Again, we see the sycamore compared with greater trees. Just as God used the lowly Amos and common sycamore to bring down the mighty Amaziah, so here He used the lesser king Hiram and the lowly sycamores to humble the mighty Solomon and his lofty cedars by making them common.

In this same chapter, there is another interesting side note that pertains to our comparison of the sycamore and sycamine trees. Verse 11 mentions that Solomon also imported *"great plenty of almug trees." (I Kings 10:11)* The almug tree, like the sycamore, has a very similar-sounding tree that is also mentioned in the Bible, the al**gum** tree. Both of these sets of trees have been thought to have one tree misspelled and to be just one kind of tree.

But most commentaries do agree that the sycamore and the sycamine are different trees. Yet they do not apply this principle to the almug and algum trees. They persist in saying that they are the same tree and that one is misspelled.

But just because two trees sound alike, this doesn't mean they are the same tree. A lot of things sound alike, but are not the same. For example, there are many Herod's in the Bible, but they are not all the same person. I believe that the words were not misspelled and that the almug and algum are two different trees, and also that the sycamore and sycamine are two different trees.

Another mention of the sycamore is in I Chronicles 27. This passage tells of the captains and officers that David set over certain portions of his kingdom. It says, *"And over the olive trees and the sycomore trees that were in the low plains was Baalhanan the Gederite." (I Chronicles 27:28)*

This speaks of *"the sycomore trees that were in the low plains."* Sycamore trees grow best in lowlands because they love water. Even today in Israel, sycamores are the mark by which lower Galilee is distinguished from upper Galilee. Where they flourish is called lower Galilee, and where they do not is called upper Galilee. They also grow in the lowlands of western Judea, along the coast of the Mediterranean Sea.

The man that was set over these trees was named Baalhanan, which means "lord of kindness." The two kinds of trees over which he was set speak of the wide range of his authority. He was over both sycamore trees and olive trees, which are very different trees. Olives flourish in dry, rocky places, but sycamores need wet lowlands. So Baalhanan didn't have a small forty-acre farm. He had a farm of many square miles, with room for great variety of land, both dry and wet, both hills and valleys.

Again in this passage, we see the sycamore named with a tree more important than itself, the olive. Just as the olive was more important than the sycamore, so David the king was more important than his lesser officer, Baalhanan.

This should be an encouragement and comfort to us. No matter how small our role may seem, it is important in God's eyes. Let us, therefore, fulfill our own individual roles with faithfulness and without growing weary in well doing.

Psalm 78 mentions the sycamore tree to describe the plagues upon Egypt. It says, *"He destroyed their vines with hail, and their sycomore trees with frost." (Psalm 78:47)* This verse refers to the seventh plague, the frost being a freezing rain that descended with the hail and destroyed the crops of Egypt.

Vines in the Bible usually refer to grape vines, though they could be others, like blackberry vines. Again, these fruitful vines are more important than sycamores, with their poor fruit. It is interesting that the sycamore is never paired with trees that are less important than itself, but always with those that are more important. The sycamore, in a unique way, shows that God is intimately concerned about things that we may consider to be insignificant.

This passage was given as a warning to Israel. God was showing Israel that just as He had punished Egypt, He could also punish Israel if she were not careful to obey Him, for God holds His Own people to an even higher standard than He does the rest of the world.

Indeed, the same warning applies to us. As Christians, we are also God's people. We have greater advantages than the world and should therefore live higher than the world. For this reason, when we do sin in spite of all these advantages, our punishment is often greater than that of the world, simply because more is expected of us. Let us learn from this warning, and strive to live up to the high standard that God expects of us.

The best known Biblical mention of the sycamore is the last one, the account of Zacchaeus. I'm sure we all remember singing the song about Zacchaeus and the sycamore tree dozens of times as children. Let us take a fresh look at this story in light of what we have learned about the sycamore.

Luke tells the familiar account, *"And Jesus entered and passed through Jericho. And, behold, there was a man named Zacchaeus, which was the chief among the publicans, and he was rich. And he sought to see Jesus who he was; and could not for the press, because he was little of stature. And he ran before, and climbed up into a sycamore tree to see him: for he was to pass that way." (Luke 19:1-4)*

Let us first notice the city in which this story took place, Jericho. Jericho was a city that was known for its abundant palm trees. It was called in the Old Testament *"Jericho, the city of palm trees." (Deuteronomy 34:3)* But the palm tree is not the tree that we remember in connection with this story. For though palm trees are tall, it is difficult to climb them. It can be done, and some people do climb palm trees, just like some people climb telephone poles. But it is not easy. Also, even if Zacchaeus had climbed a palm tree, he couldn't have sat on the fronds. They are too light and would have broken under the weight of a man.

The sycamore, on the other hand, was much easier to climb. It must be remembered that the Biblical sycamore was not our North American sycamore with branches high off the ground. The Biblical sycamore had branches much lower to the ground. Zacchaeus could also eat the sycamore fruit while he waited. Thus it was an ideal tree for him to sit in and wait for Jesus.

Again, in this account, we see a comparison of importance. The palm tree is more important in terms of usefulness than the sycamore tree. And of course, Jesus Christ, Who is of supreme importance, is far more important than Zacchaeus.

We also see another similarity between Zacchaeus and the sycamore. Just as Zacchaeus was low in stature, so the sycamore tree was lower in stature than the palms with which it was surrounded.

This is symbolic, not only of Zacchaeus' low physical stature, but also of his low standing in society. He was a publican, and not only that, but *"chief among the publicans." (Luke 19:2)* As such, he was well known as a thief and robber, and was held in contempt by all. Thus, both literally because of his size and figuratively because of his position, he was a man upon whom everyone looked down.

But Zacchaeus did the right thing with his low position. He did not try to lift himself up with his own righteousness. Instead, he took a step of humility by climbing this sycamore tree. For one thing that is often not considered in this story is the fact that

Zacchaeus was probably not the only one to climb this tree. Instead, many little children probably also climbed the sycamore tree in order to see Jesus. They, like Zacchaeus, were short and couldn't see Jesus over the tall adults about them.

It was probably a very humbling and perhaps embarrassing thing for Zacchaeus, a grown man, to climb this tree with all these children. But Zacchaeus wasn't ashamed. He wanted to see Jesus, and he knew this was the only way that he could, by humbling himself as a little child.

Jesus said in another place that all men must come to Him this way. He said, *"Verily I say unto you, Whosoever shall not receive the kingdom of God as a little child, he shall not enter therein." (Mark 10:15)*

Anyone who wants to come to Jesus cannot come proclaiming himself to be important. He cannot come as a man of great stature. Instead he must, like the sycamore tree, take the place of least importance. And he must, like Zacchaeus, come as a little child. For all who come this way, Jesus will say, as He said to Zacchaeus, *"This day is salvation come to this house." (Luke 19:9)*

This has been the theme of the sycamore tree, that God loves the humble. It is the humble souls that He receives into His kingdom and that He uses in His service.

43
The Teil Tree

The teil tree is named only once in the Bible, in Isaiah 6:13 which says, *"But yet in it shall be a tenth, and it shall return, and shall be eaten: as a teil tree, and as an oak, whose substance is in them, when they cast their leaves: so the holy seed shall be the substance thereof."* There is much debate over what this tree is. Many commentaries say it is a terebinth tree, others say it is an oak, and others an elm.

Those who say the teil is a terebinth argue that the teil does not grow in Israel, whereas the terebinth does. But the teil probably would have grown in the lush Israel of Isaiah's day, 2,700 years ago, and it probably still would grow there in certain areas if it were planted there.

The English word "terebinth" is a relatively new term. Though it is found in the apocrypha, the translators did not use it in the canonical books of the King James Version of the Bible.

The properties of the terebinth are such that it is improbable that the translators would have confused the it with the teil tree, or even with the oak or elm. The terebinth doesn't look anything like these trees or share properties with them. It is a pistachio tree and bears berries. For all these reasons, I do not believe the teil tree to be a terebinth tree.

Those who say the teil tree is really an oak are also faced with a problem, for the teil is mentioned side by side with the oak. If it were an oak, this would make the verse read, "as an oak tree, and as an oak," which would make no sense.

The elm tree has already been mentioned in another place in the Bible, and I do not believe the same tree would have been translated two different ways. So we can rule out the elm for the teil as well.

The Hebrew word translated "teil" is a word we have seen before, the word *elah* (אלה) which literally means "strength." This is the same word translated "elms" in Hosea 4:13, and it is also one of the words translated "oak" in the Bible. Its root meaning of strength is the key to interpreting it. It does not

Carpenters know the difference between the western and eastern redcedars because their wood is different and is used for different purposes. When we go to buy a standard cedar board, it is western redcedar. This wood is brownish with tan in it. But if we go around the corner and buy a specialty piece of cedar for a chest of drawers or a closet, it is eastern redcedar, which is reddish with tan. Both are aromatic, and only the color difference reveals the kind.

In order to distinguish between the two trees while they are living, we must look at their cones. Their leaves are about the same, but their cones have some definite differences. The cones of the eastern redcedar, the *Juniperus*, have scales that are fused together. Because of this and their small size, the cones look like berries. But the cones of the western redcedar, the *Thuju*, are not fused together and look like the familiar pinecones, or probably more like spruce cones because of their small size.

But I do not believe the trees of the genus *Thuja* to be the proper interpretation of thyine wood. I think more probable is the second suggestion, the genus *Thyia*. This wood is known as *Thyia tetraclinis*, called by the ancient Romans citron, and is also in the cypress family. But it is a more precious wood than the wood of the *Thuja* genus. It is probably most similar in its characteristics to the arborvitae, but much more precious. Its wood has a distinctive cedar smell and is hard and oily.

We must keep in mind the context of the passage. The merchants are mourning because no one will buy their precious merchandise anymore. This means they will be reduced to poverty, which causes them to weep. So in order to be mentioned with gold and silver, thyine wood must be very precious.

It is also interesting to note that thyine wood is mentioned separately from *"vessels of most precious wood."* This is another factor that causes me to think it is *Thyia tetraclinis*. To be mentioned separately even from *"most precious wood,"* thyine wood must be extremely precious.

Indeed *Thyia tetraclinis* fits this qualification. Even a small piece of its wood, two inches by six inches by six inches, costs one hundred dollars. This wood is very precious not only because of its useful qualities, but also because of its unusual and beautiful aesthetic qualities. It does not have a pronounced grain like most wood, but is a speckled wood that looks somewhat similar to bird's eye maple. A person would use only a little of this wood and would use it only for very special things, not even for vessels. There are records that this wood was used once to make a violin, no doubt an extremely expensive violin.

This kind of tree grows in the same places and conditions as juniper trees. But junipers are the more resilient of the two. This means that whenever these trees are cut down, they don't repopulate themselves easily, and so junipers fill in the vacancy and take over the area.

This characteristic further adds to the rarity and expensiveness of thyine wood, for the most precious wood is taken from the root. This means that the *Thyia tetraclinis* tree must be dug up and destroyed. Once this is done, a new tree is very hard to cultivate, making the price of the wood very high.

The merchants of this passage furnish us with a sobering warning against materialism. The Bible says of them, *"The merchants of the earth are waxed rich through the abundance of her delicacies." (Revelation 18:3)* These merchants are slaves to things, things that bring gain and pleasure to them.

These merchants are the opposite of the Israelites of Moses' day. The Israelites brought all their precious things to Moses and gave them to the Lord for use in the tabernacle. Instead of making themselves slaves to their material goods, they willingly made themselves slaves and servants to Jesus Christ.

But the merchants of Revelation have made themselves slaves to their own gain and pleasure. Now they weep because their gain is gone, and they have lost everything. The Bible says, *"thou shalt find them no more at all." (Revelation 18:14)* This passage goes on to show just how fleeting material treasures are, for they are all lost in just one brief hour. It says, *"For in one hour so great riches is come to nought." (Revelation 18:17)* Because of their slavery to such temporal, material goods, *"The merchants of these things, which were made rich by her, shall stand afar off for the fear of her torment, weeping and wailing." (Revelation 18:15)*

Thyine wood warns us of the grave danger of becoming enslaved to material goods. It is very rare and fleeting itself, being easily subject to destruction through harvesting and marketing. Also its difficulty in repopulating itself speaks of the fact that once material goods are gone, they are never to be recovered. *"Thou shalt find them no more at all." (Revelation 18:14)* Let us be warned and not become enslaved to the things of this world, but rather let us dedicate ourselves and all our goods as servants of the Lord.

46
The Tree of Life in Genesis

The Tree of Life

Looking at tree of life, we must discuss one question that has been debated for centuries. Did Adam and Eve eat of the tree of life before they ate of the tree of the knowledge of good and evil?

This is a somewhat difficult question to address. We must be very careful not to overanalyze because there is a danger of reading in something that is not there and also of taking away something that is there.

For example, over the years the fruit that Adam and Eve ate has been overanalyzed. People have speculated that it was an apple, but in reality we don't know. The Bible doesn't tell us because it is not important for us to know. But what is clear in the Bible and what is important is the issue
of obedience. Obedience is the thing that the tree of the knowledge of good and evil teaches us, not what kind of fruit was on it. When we overanalyze the fruit, we miss the important point, obedience.

In this chapter, I am seeking to deal with the tree of life separately from the tree of the knowledge of good and evil. This is somewhat difficult because they are in the same garden and mentioned in the same place in the Bible. In fact some even say they are the same tree. However from the clear Bible language this cannot be true.

I also hope to deal with the tree of life in Genesis separately from the tree of life in Revelation. This is even more difficult because I do believe these to be the same tree. More detail is given in Revelation because the tree has additional aspects and applications from what it has here in Genesis.

We first look at Isaiah 53:2. It prophesies of Christ, *"For he shall grow up before him as a tender plant, and as a root out of a dry ground: he hath no form nor comeliness; and when we shall see him, there is no beauty that we should desire him."*

Here Christ is likened to an unattractive plant. As the tree of life is a picture of Jesus Christ, perhaps this tree was also unattractive to the eye. Christ is not attractive to the world simply by His looks. As we saw in the chapter 20 on the garden of Eden, men must taste of Him in order to see that He is good.

Galatians 5:22-23 also gives some insight. *"But the fruit of the Spirit is love, joy, peace, longsuffering, gentleness, goodness, faith, Meekness, temperance: against such there is no law."* These could be part of the fruit of the tree of life or byproducts of its fruit.

The location of the tree of life is also significant. It was *"in the midst of the garden." (Genesis 2:9)* Yet perhaps surprisingly, the tree of the knowledge of good and evil was also *"in the midst of the garden." (Genesis 3:3)* The garden here symbolizes our lives, and the two trees symbolize the occupants of our hearts. All of us begin life with the knowledge of good and evil in the midst of our hearts. It is sin that reigns within us and controls us. Yet at salvation Jesus Christ comes to dwell within our lives Himself. Just as the tree of life was in the midst of the garden, Jesus Christ should be in the midst of our lives. He should be the very center of our lives so He can be the one to reign over and control us.

Adam was made from the dust of the ground. Was it God's purpose for him that he live forever? Was he to be immortal? After the fall, God did say, *"for dust thou art, and unto dust shalt thou return." (Genesis 3:19)* But was it so from the beginning?

One thing is clear, whether Adam was intended to be immortal or not, whether he ate of the tree of life before his fall or not, he had the opportunity to take of the tree of eternal life. Acts 5:30 says, *"The God of our fathers raised up Jesus, whom ye slew and hanged on a tree."* This tree, the cross of Christ, was in the midst of the trees of the two thieves, one on His right and one on His left. The tree in the midst, the cross of Christ, is in a very real sense the tree of life, the tree of eternal life. To us is offered eternal life because Jesus Christ hung on that tree. Without that tree, planted on Golgotha Hill, we would not have eternal life.

The tree of life, the picture of Jesus Christ, spans the entire Bible, from Genesis to Revelation. Only in Revelation is it again called *the* tree of life. I believe it is the same tree because God preserves all things that are holy. He preserves our justified lives, made holy through the righteousness of Christ, and He preserves His holy Word. He said, *"Heaven and earth shall pass away, but my words shall not pass away." (Matthew 24:35)* He even places His Word above His name by binding Himself never to violate His Word. Even if He wanted to change His own mind, He couldn't because He has bound Himself to His holy, eternal Word. He also preserves His holy Son, Jesus Christ.

This Jesus Christ is the very picture of the tree of life. Jesus said, *"I am the way, the truth, and the life: no man cometh unto the Father, but by me." (John 14:6)* Like *the* tree, Jesus is *the* way, *the* truth, and *the* life.

Psalm 34:8 invites us to taste of this tree. *"O taste and see that the LORD is good: blessed is the man that trusteth in him."* He may not be appetizing to the eyes. Like a food that is unattractive, we may not naturally say that He looks good. We may even avoid Him. But if we will taste, we will find that He is good, both good to our taste and good for us.

We now address the question that has been debated for centuries. Did Adam and Eve eat of the tree of life before they ate of the tree of the knowledge of good and evil? The Bible doesn't say whether they did or did not. It does make it clear that they did not eat of it *after* eating of the tree of the knowledge of good and evil.

One argument says that Adam and Eve would have had no time to eat of the tree of life. The literal six-day Creation, which I believe, seems to indicate a relatively short time that Adam and Eve lived in the garden before the Fall.

On the other hand, we do know that Adam had the time to name the animals. But we don't know how long this took. For me it would probably have taken years. But for Adam, with his superior brain power, it may have taken only a few hours.

We are not told how long this took. Neither are we told if Adam and Eve ate of the tree of life during this time prior to the Fall. The Bible does not say they did, and it does not say they didn't.

Another argument for them not eating of the tree of life prior to eating of the tree of the knowledge of good and evil is that the garden may have been so big that they didn't get to the tree of life. This is a possibility. If it were unattractive and undesirable, they may have held off from eating it until a later time.

Yet another argument is that they could not have eaten of the tree of life, or else they would have lived forever physically. This is perhaps plausible, if we consider the symbolism. For one to come to Christ for salvation, he needs to come only once. He doesn't need to come again.

Yet we may also come to Christ in prayer time and time again. Perhaps eating of the tree of life is not likened to coming to Christ for salvation, which is done only once, but to coming to Him in prayer, which may be done repeatedly. In this case, one taste would not have produced eternal physical life, but instead a fellowship and communion that was then broken when Adam and Eve sinned.

A final argument against them eating of the tree of life before the tree of the knowledge of good and evil is that if they had eaten, they would never have sinned. They would have had no desire to know good and evil, because of the knowledge that they had. This is not necessarily true. They would not necessarily have received knowledge from the tree of life.

While these arguments are not conclusive, one thing is clear; regardless of the answer to this question, one who comes to Jesus Christ for salvation needs to come only once. Romans 6:22-23 says, *"But now being made free from sin, and become servants to God, ye have your fruit unto holiness, and the end everlasting life. For the wages of sin is death; but the gift of God is eternal life through Jesus Christ our Lord."* The question about Adam and Eve eating of the tree of life, however, may not be restricted just to salvation, in which case they could eat more than once.

God promised, *"But of the tree of the knowledge of good and evil, thou shalt not eat of it: for in the day that thou eatest thereof thou shalt surely die."* (Genesis 2:17) Whether or not they ate of the tree of life, this promise must be fulfilled. They must die after eating of the tree of the knowledge of good and evil. God's commandment and His punishment stand regardless of what men do. What He says He will do, He will do because He doesn't go back on His Word.

The Bible doesn't mention that Adam and Eve did not eat of the tree of life. So it is possible that they did. But neither does it mention that they did eat of it. And again it is possible that they didn't. I will say only that it looks like they could have.

I Corinthians 15 teaches us something of our access to Jesus Christ. *"And so it is written, The first man Adam was made a living soul; the last Adam was made a quickening spirit. Howbeit that was not first which is spiritual, but that which is natural; and afterward that which is spiritual. The first man is of the earth, earthy: the second man is the Lord from heaven. As is the earthy, such are they also that are earthy: and as is the heavenly, such are they also that are heavenly. And as we have borne the image of the earthy, we shall also bear the image of the heavenly. Now this I say, brethren, that flesh and blood cannot inherit the kingdom of God; neither doth corruption inherit incorruption."* (I Corinthians 15:45-50)

Access to the tree of life, to Jesus Christ, was denied when Adam and Eve were cast out of Eden after they sinned. The first Adam, as I Corinthians states, was made of flesh and blood. The second Adam was a quickening spirit. The application for us is clear.

While we have no access to the physical tree, we do have the more important access to Jesus Christ Himself. We have access not to a mere tree, a picture of Christ, but to Himself, the quickening spirit.

This access that we enjoy was also no doubt enjoyed by Adam and Eve before they sinned. Whether they ate of the physical tree or not, the Bible pictures a fellowship and communion between them and the Lord before their sin.

Throughout the Bible, there are many examples of this fellowship with Christ, often pictured by eating together with Him. I Corinthians 11 gives the account of the Lord's Table, or as it also called Communion, picturing our communion with Him. *"For I have received of the Lord that which also I delivered unto you, That the Lord Jesus the same night in which he was betrayed took bread: And when he had given thanks, he brake it, and said,*

47
The Tree of Life in Revelation

We now look at the tree of life as it appears in Revelation. More detail about this tree is given here than in Genesis. But we do have good reason to believe that it is the same tree. In both places it is called **the** tree of life, indicating that both references are different glimpses of the same tree.

The tree of life appears in Revelation in chapter 22. *"And he shewed me a pure river of water of life, clear as crystal, proceeding out of the throne of God and of the Lamb. In the midst of the street of it, and on either side of the river, was there **the tree of life**, which bare twelve manner of fruits, and yielded her fruit every month: and the leaves of the tree were for the healing of the nations." (Revelation 22:1-2)*

The first thing we notice about this tree is that the Bible doesn't say it grows in soil. Not all trees and plants need to grow in soil. For example, there are hydroponic trees, trees that are grown entirely in nutrient-filled water and not in soil.

But what is soil? It is made up of four basic components, sand, silt, clay, and humus. These can be reduced even further to two basic ingredients, decayed organic matter and weathered rock. Humus is decayed plant and animal matter. The sand, silt, and clay are all rock particles, weathered by either chemical or mechanical means. Thus soil is merely the collection of these particles.

With this definition of soil, no plant really grows **in** soil. I used to ask my college students, "Where do roots grow?" Of course, their first answer would be, "In soil." But the answer for which I was looking was, "In the air pockets between soil particles." Plants receive nutrients from the soil, but they do not actually grow **into** the soil. Thus, even in technical terms, the tree of life doesn't have to grow in soil.

This tree instead grows in pure water. *"And he shewed me a pure river of water of life, clear as crystal, proceeding out of the throne of God and of the Lamb." (Revelation 22:1)* This water is perfectly pure. Here on earth, we have never seen truly pure water such as this. Throughout the years, man has tried various ways of making pure water. For example, there is a purified water called light water. It is the opposite of heavy water, which is used in nuclear reactors. Heavy water has deuterium, an isotope of hydrogen, added to it. Light water, on the other hand, is depleted of deuterium. This makes it very low in toxicity and also in electrical conductivity. This light water is used in the medical field for its healing affect on people with cancer and other diseases, and also because it slows the process of aging.

Though light water is not perfectly pure, it demonstrates the truth that water does have degrees of purity, and that it is theoretically possible to have perfectly pure water. It also shows that the more pure water is, the more health benefits it has.

Another interesting property of this water is found in chapter 21. *". . . and the city was pure gold, like unto clear glass . . . and the street of the city was pure gold, as it were transparent glass."* *(Revelation 21:18, 21)* The city itself through which this water flowed was pure gold. Studies have shown that if water is infused with a little gold and silver, it has even added health benefits.

These health benefits were seen in the medieval days. In these days, the nobility would drink from gold and silver vessels, while the peasants would drink from vessels of wood or lead. Often the noblemen were healthier than the common folk. Though their drinking vessels were certainly not the only difference between the two classes that would affect their health, it no doubt was a contributing factor.

The tree of life thus has the ideal location. It is located *"in the midst of the street of it, and on either side of the river."* *(Revelation 22:2)* Here it can draw upon the pure water of life, and receive the benefits of the street of gold. We know from Genesis that eating of this tree of life makes man live forever. *"And the LORD God said, Behold, the man is become as one of us, to know good and evil: and now, lest he put forth his hand, and take also of the tree of life, and eat, and live for ever."* *(Genesis 3:22)* Perhaps it is the health benefits of pure water infused with gold that make this tree such a life-giving tree. Drawing upon the water of life, this tree can only be the tree of life. It has life as both the sustenance upon which it draws and the sustenance it offers to others.

Life is the dominant theme of this passage in Revelation. Chapter 21 mentions the *"book of life." (Revelation 21:27)* Chapter 22 mentions both *"water of life" (Revelation 22:1)* and *"the tree of life." (Revelation 22:2)* This abundance of life confirms the truth that there will be no death in heaven. There is no place for death in the presence of all this life. Christ has decreed it so, establishing life, not death, in heaven.

Another characteristic of the water of life is its clarity. It is described as *"clear as crystal." (Revelation 22:1)* This is the same description given of the sea of glass before God's throne, from which this river seems to flow. *"And before the throne there was a sea of glass like unto crystal." (Revelation 4:6)* This purity of this water of life gives it a clear reflection like that of a mirror. It reflects honestly whatever image that comes before us. Its purity reveals us just as we are, and reveals the holiness and glory of Jesus Christ as He is.

Many of us have seen the waves of an earthly sea sparkle. The light reflected from water can be almost blinding at times. But imagine the brilliant light that will be when the water that reflects is pure water and the glory that is reflected is the glory of Jesus Christ. Such a sight will be glorious to behold.

Zechariah also speaks of living waters flowing from God's presence. He speaks of the day of the LORD, saying, *"And it shall be in that day, that living waters shall go out from Jerusalem; half of them toward the former sea, and half of them toward the hinder sea: in summer and in winter shall it be." (Zechariah 14:8)* The place of God's habitation is the center from which living waters flow.

We now turn our focus to the tree of life itself. In order to understand this tree fully, we will be looking often at Ezekiel 46 and 47. These chapters have many parallels with the passage in Revelation and give us a deeper understanding of it. For example, Revelation 21 introduces chapter 22 by describing the city of new Jerusalem. It says of the city, *"And the gates of it shall not be shut at all by day: for there shall be no night there." (Revelation 21:25)* Ezekiel 46 gives a contrast with earthly Jerusalem. *"Thus saith the Lord GOD; The gate of the inner court that looketh toward the east shall be shut the six working days; but on the sabbath it shall be opened, and in the day of the new moon it shall be opened." (Ezekiel 46:1)* For this reason, we will look often to Ezekiel as we study the tree of life.

The tree of life is first introduced in Revelation with these words, *"In the midst of the street of it, and on either side of the river, was there the tree of life." (Revelation 22:2)* How is it possible for the tree to be both in the midst of the street and on either side of the river?

To answer this question, we first look at Ezekiel 47. *"And by the river upon the bank thereof, on this side and on that side, shall grow all trees for meat." (Ezekiel 47:12)* This is a similar description to that in Revelation, but seemingly not the same. In Ezekiel it is several **_trees_** that grow on both sides of the river, while in Revelation it is **_the tree_**.

Some say that the tree of life is a figurative term for what is actually many trees, like the trees in Ezekiel. We will take the literal interpretation of the singular term and consider the tree of life as one tree.

However, it is certainly possible for it to seem like many trees. The banyan tree, which we mentioned in chapter 17, is a good example of this. One banyan tree can cover several acres. This is because the banyan puts down roots from its branches to the ground. Each new root looks like a separate tree trunk. But in reality the "trunks" are all interconnected and are just the roots of one tree. It is interesting to note that the banyan tree is in the *Ficus* family, the same family as the fig tree, which is prominent among the trees of the Bible. It is possible that the tree of life is similar to the banyan tree.

Another kind of tree that can look like several trees is the aspen. In the Wasatch Range of the Rocky Mountains, there is a particularly notable aspen that we also mentioned in chapter 33 of this book. This aspen appears to be a network of 47,000 individual aspen trees. But on closer examination, we find that all 47,000 "trees" are growing from a single tree root. Since they are a single tree, they all change color and drop their leaves at

entire history, it has never had such numbers. Also the land promised to Abraham covers a far greater area than Israel has ever possessed. These hanging promises will eventually be fulfilled in the day when Jew and Gentile will be one in Christ, when they will all come together in the new heaven, the new earth, and the new Jerusalem. This is the healing of the tree of life, the restoration and unity of these entire nations in Christ for all eternity. Such is the greatness, the magnitude, and the abundance of life of this tree.

To conclude this look at the tree of life, we address one final question that arises concerning it. Some believe that it is merely a figurative symbol and not a real physical tree. On the surface, it seems that this could possibly be so. The tree of life pictures Christ and His Word. This is certainly symbolic, for we don't actually consume Christ or eat the Bible. We consume them spiritually by taking them into our hearts and lives.

But this view leads to many difficulties. First, if the tree of life is figurative, then the tree of the knowledge of good and evil must also be figurative. Indeed, many who hold to the first hold to the second as well. But where do we stop? Maybe the garden of Eden was not real. Maybe Adam and Eve were not real. Such reasoning will soon lead to the whole Bible being figurative. Thus to call the tree of life figurative is to put the whole Bible on shaky footing and to subject it to our own reasoning and interpretation, narrowing its application.

In Genesis 2:9 the Bible says clearly that God made both the tree of life and the tree of the knowledge of good and evil to grow. *"And out of the ground made the LORD God to grow every tree that is pleasant to the sight, and good for food; the tree of life also in the midst of the garden, and the tree of knowledge of good and evil."* (Genesis 2:9) How could God make something to grow from the ground that was not real, but only figurative? Just taking the Bible literally, it indicates that the tree of life is a real tree.

We must note, however, that there are things in the Bible that are figurative. For example, Christ says, *"Behold, I stand at the door, and knock: if any man hear my voice, and open the door, I will come in to him, and will sup with him, and he with me."*

(Revelation 3:20) We know that Christ doesn't literally stand outside of a wooden door and knock. The Bible also says, *"the earth . . . is his footstool." (Matthew 5:35)* We know that the earth is not literally a three-legged stool where the Lord rests His feet. But when something in the Bible is figurative as these are, it is usually obviously figurative. Figures and symbols in the Bible are used to help us grasp things we cannot otherwise understand.

Also in this book, a lot of what we discuss is figurative, especially in chapter 26 on human attributes given to trees. For example, the fig tree in chapter 26 represents the nation of Israel. But that doesn't mean it is not also a real tree. The trees used to build the temple also represent certain things, but if they were not real, then the temple constructed from them was not real. Figurative meaning does not exclude real existence. Real trees can and do picture figurative things.

The phrase *"**a** tree of life"* is always symbolic. It is used to picture wisdom, the fruit of the righteous, fulfilled desire, and a wholesome tongue. But ***the*** tree of life is a literal tree, grown from the ground in the garden of Eden, and living throughout eternity. This distinction between ***the*** and ***a*** is even seen in the Hebrew letters. In Hebrew, ***a*** tree of life is Myyx Ue (Etz Chayim). But ***the*** tree of life is Myyxh Ue (Etz HaChayim). The added syllable h (Ha) is the definite article, which applies to the entire phrase, making it read "the tree of life."

The tree of life is a real tree, and at the same time, it pictures many things for us. Real and living, spanning from Genesis to Revelation, it stands as the emblem of eternal life from eternity past to eternity future. The life found in it is freely offered to all through the person of Jesus Christ. It is He Who is pure water, never decaying fruit, and perfectly healing leaves. It is He Who is ultimately pictured in the tree of life.

In order to see this beautifully illustrated, let us close this chapter by seeing how Jesus Christ is pictured as different parts of a tree throughout the Bible. He is first the **seed** in His humanity. He was born of the seed of David according to John 7:42, Romans 1:3, and II Timothy 2:8. And He took on Himself the seed of Abraham according to Hebrews 2:16.

Christ is next the **root**. He is called the root of David in Revelation 5:5 and 22:16. He is also called the root of Jesse in Isaiah 11:10 and Romans 15:12.

He is also the **stem**, or the **vine**. He is called the stem of Jesse in Isaiah 11:1, and He calls Himself "the vine" in John 15:5.

He is further the **branch**. He is called the branch in Zechariah 6:12 and Isaiah 11:1. He is the righteous branch in Jeremiah 23:5 and the branch of righteousness in Jeremiah 33:15. He is also the servant branch in Zechariah 3:8.

And finally, Jesus Christ is the **fruit**. Being the same essence as the Holy Spirit, He is the firstfruits of the Spirit, as described in Romans 8:23. And He is also the firstfruits of I Corinthians 15:23.

Truly Jesus Christ is the perfect, everlasting, life-giving tree of life. Let us partake of Him by faith and thereby receive the everlasting life that He offers.

48
The Tree of the Knowledge of Good and Evil

We now turn to examine the other named tree in the garden of Eden. This is the tree of the knowledge of good and evil. Because the garden is now hidden from our eyes, we cannot look upon it and upon the trees in it with our physical eyes. Therefore we must look upon them with spiritual eyes.

As we said for the tree of life, there are no two trees alike. Trees can even be cloned, but there are still differences in shade, sunlight, water, nutrients. Because of these differences, even two otherwise identical trees, one cloned from the other, will exhibit different growth patterns. They will still not be exactly alike.

The tree of the knowledge of good and evil, like the tree of life, is a unique tree. Like the tree of life, it can now be seen only with spiritual eyes, not with physical eyes.

Genesis 3 opens with a dialogue between Eve and Satan concerning the tree of the knowledge of good and evil. *"Now the serpent was more subtil than any beast of the field which the LORD God had made. And he said unto the woman, Yea, hath God said, Ye shall not eat of every tree of the garden?" (Genesis 3:1)*

With this question, Satan very subtly planted a doubt in Eve's mind, for this is how Satan works. Here he simply questioned God, *"hath God said . . .?"* Did God really say this? Or is it open to interpretation?

These questions, subtle though they be, are a direct contradiction to the Word of God. God's Word teaches that what God says is absolute and that *"no prophecy of the scripture is of any private interpretation." (II Peter 1:20)*

Satan would have us believe that there are gray areas. In reality, there is no gray area. Everything is either black or white. The reason people seek gray areas is because they have some sin to which they are clinging. They cannot find that particular sin named in the Bible, though if they would study it out, they would

Solomon had knowledge. He executed Joab and Shimei, two men who had been thorns in the side of David.

But even this was not wisdom. Solomon confessed, *"I am but a little child: I know not how to go out or come in." (I Kings 3:7)* This didn't mean that he was dumb. He had some knowledge about the kingdom, but he lacked the wisdom to put that knowledge into practice and rule rightly. Recognizing the difference between the two, he requested yet more knowledge and also wisdom to apply that knowledge in II Chronicles 1:10. *"Give me now wisdom and knowledge, that I may go out and come in before this people: for who can judge this thy people, that is so great?" (II Chronicles 1:10)*

In Colossians 1:9, Paul also recognized this difference between knowledge and wisdom and prayed for both of them. *"For this cause we also, since the day we heard it, do not cease to pray for you, and to desire that ye might be filled with the **knowledge** of his will in all **wisdom** and spiritual understanding." (Colossians 1:9)*

But Eve mistook wisdom and knowledge to be the same thing. Thinking she was getting wisdom, she reached out and took of the fruit of the tree of the knowledge of good and evil.

Why did God put this tree in the garden? Why did He give Adam and Eve a choice of obedience or disobedience? Why did He give them an opportunity to sin?

This is the way God works. God always gives people a will. We must make the choice whether to serve Him or not to serve Him. God didn't make robots. He wants us to choose to love Him voluntarily.

But when we choose sin, that sin will leave its mark upon us. An example of the devastating effect of sin is found in the book of Ruth.

In this book, Elimelech and Naomi sinned by disobeying God and leaving the land of Israel. They took their family and left Bethlehem, the house of bread, to escape famine. They came to Moab, and there Elimelech and his two sons, Mahlon and Chilion, died. When Naomi returned widowed and childless to Bethlehem, sin had left a noticeable mark upon her. Even her own neighbors had trouble recognizing her. As she and her daughter-in-law Ruth entered the city, the Bible says, *"And it came to pass, when they were come to Bethlehem, that all the city was moved about them, and they said, Is this Naomi?" (Ruth 1:19)*

Even Naomi recognized the change in herself. She answered, *"Call me not Naomi, call me Mara: for the Almighty hath dealt very bitterly with me." (Ruth 1:20)* Sin mars the life so that it becomes hard to recognize.

For this reason, I believe that Adam and Eve looked different after eating of the tree of the knowledge of good and evil. Even today we can see the effects of sin on people's lives. Drugs cause people to age faster. Drinking and smoking both leave their marks. Sin affects one's appearance, and I believe it affected the appearance of Adam and Eve.

Also because of sin, a curse was brought upon the earth. The ground, Adam, Eve, and the serpent were all cursed. This evil was a direct result of eating of the tree of the knowledge of good and evil. The tree and the fruit itself were not evil. But eating of the fruit had the ability to reveal evil. That is why it was called the tree of the ***knowledge*** of good and evil.

Until they ate of this tree, Adam and Eve had knowledge only of good. Now their eyes were suddenly opened, and they had knowledge of evil. It is like a person born blind or deaf. If suddenly, through surgery or some other means, that person's eyes or ears were opened, his world would be so very new and different.

Through eating of this tree of the knowledge of good and evil, Adam and Eve were saying in essence, "I will determine what is good and what is evil." They chose to gain knowledge and determine right and wrong on their own. People are still saying that today. But that is not what Jesus Christ says. He determines what is good and evil. We should simply obey.

Because of their disobedience, the curses fell upon Adam and Eve. For Adam, he must till the ground in sorrow, sweat, and toil. For Eve, she must have pain in childbirth, and her desire would be to her husband. This phrase means literally that she would desire to be the husband, the head of the home. For the serpent, he must go upon his belly and eat dust all the days of his life. Even today, the serpent is one of the most repulsive of all the animals on the earth. It is not so much his looks that repels people, but his sneakiness.

The word for curse also contains the idea of confusion or chaos. Adam would be in chaos, having a duty to till the ground, yet having to toil with thorns and thistles. Eve would be in chaos, trying to reverse the roles of husband and wife. The serpent would be in chaos, going from the most beautiful angel, Lucifer, to the most repulsive animal. Even the ground would be in chaos. Flowers would be planted, but thorns and thistles would grow up also.

Even in this cursing, God's sovereignty can be seen. God wants man to work. He doesn't want him to be idle, to sit around with nothing to do. Therefore he cursed the ground in order that man might spend his days in honest work.

The tree of the knowledge of good and evil has left us with a sobering lesson. We don't know what it looked or tasted like. That is not important. We know that it was different from the tree of life and that there is no other tree like it, for there are no two trees exactly alike. We also know that only the tree of life is recorded as being in Heaven. The tree of the knowledge of good evil seems to have ceased to exist. But in its vanishing from existence, this tree has left us something of far more importance than its looks or taste. It has reminded us that God has given us a choice, obedience or disobedience. Its silent testimony witnesses of the sobering consequences of disobedience, and impresses upon us the importance of whole-hearted obedience to Jesus Christ.

49
Tree Removal

We now look at Biblical examples of felling trees. Often the felling of trees is used figuratively to speak of destroying a large group of people. Isaiah prophesied of such destruction upon the vast Assyrian army. *"Behold, the Lord, the LORD of hosts, shall lop the bough with terror: and the high ones of stature shall be hewn down, and the haughty shall be humbled. And he shall cut down the thickets of the forest with iron, and Lebanon shall fall by a mighty one." (Isaiah 10:33-34)*

This prophecy was fulfilled in the reign of Hezekiah. Sennacherib, the king of Assyria, sent his general Rabshakeh *"with a great host against Jerusalem." (II Kings 18:17)* The Assyrian army was thick like trees. As they encamped about Jerusalem, they covered the ground like a forest.

But Godly king Hezekiah prayed to the Lord for deliverance from the great army. The next day the Lord cut down the dense thicket of men, just as Isaiah had prophesied. *"And it came to pass that night, that the angel of the LORD went out, and smote in the camp of the Assyrians an hundred fourscore and five thousand: and when they arose early in the morning, behold, they were all dead corpses." (II Kings 19:35)* God Himself is a tree feller, felling here an entire army.

Psalm 74:5 is an interesting verse on the felling of trees. *"A man was famous according as he had lifted up axes upon the thick trees." (Psalm 74:5)* Perhaps one fulfillment is the Lord Himself Who lifted up His axe upon the forest of the Assyrian army.

Another fulfillment may be Gideon. The Hebrew name *Gideon* means tree feller. Indeed Gideon did become famous for his tree felling. He not only literally cut down the grove by his father's altar to Baal, but he also cut down the grove of the invading Midianite army.

Another famous tree feller was Hiram, the king of Tyre. He became famous as the feller of trees for Solomon's temple, supplying him with cedar, fir, and almug trees.

One day Christ will also fulfill this verse. In one great battle at the end of time, He will fell all the trees that yet stand against Him. There at the battle of Armageddon He will wipe out millions of men at one time, like felling a forest of millions of trees.

Yet today Christ has the power to fell His enemies. For those stand against Christ, against His chosen people of Israel, or against Jerusalem, no matter how great their army is, Christ will fell them. But the opposite is also true. For us who are on the side of Christ, we do not need a big army or mighty warriors in order to obtain victory. We need only to stand with Christ. Alone, He can fell the largest army that can form against us.

Another occasion when the Bible uses the felling of trees as an example is Nebuchadnezzar's invasion of Egypt. In Jeremiah 46, this invasion is described. Egypt had thousands of cities and is described in this passage as a forest. *"The voice thereof shall go like a serpent; for they shall march with an army, and come against her with axes, as hewers of wood. They shall cut down her forest, saith the LORD, though it cannot be searched; because they are more than the grasshoppers, and are innumerable." (Jeremiah 46:22-23)*

In order to defeat Egypt, Nebuchadnezzar had to destroy all these cities. Thus he came with an army as numerous as a forest of trees. These warriors came as hewers of wood, ready to fell the thousands of Egyptian cities. Just as Jeremiah had prophesied, Nebuchadnezzar and his army cut down the cities like trees and destroyed them.

A final use of tree felling as an example is in Joshua 17. Here the children of Israel had entered the land, and the individual tribes were conquering their portions. *"And Joshua answered them, If thou be a great people, then get thee up to the wood country, and cut down for thyself there in the land of the Perizzites and of the giants, if mount Ephraim be too narrow for thee." (Joshua 17:15)*

The tribes of Ephraim and Manasseh were complaining to Joshua that they did not have enough land. *"And the children of Joseph said, The hill is not enough for us: and all the Canaanites that dwell in the land of the valley have chariots of iron, both they who are of Bethshean and her towns, and they who are of the valley of Jezreel." (Joshua 17:16)*

But Joshua answered them that their land was plenteous but wooded. The trouble was that they hadn't yet cut down the trees in their own land. *"And Joshua spake unto the house of Joseph, even to Ephraim and to Manasseh, saying, Thou art a great people, and hast great power: thou shalt not have one lot only: But the mountain shall be thine; for it is a wood, and thou shalt cut it down: and the outgoings of it shall be thine: for thou shalt drive out the Canaanites, though they have iron chariots, and though they be strong." (Joshua 17:17-18)* Ephraim and Manasseh had to remove the trees in the land they already possessed before they could conquer new land.

In the same way, we must cut down the forests in our own lives before God will give us more. We should not ask for more responsibility or more blessing if we haven't fully used what God has already given us. We must fell the trees that keep us from fulfilling the responsibilities we already have, and from serving Christ. We must fell the trees that block out the sunlight of Christ and His Word. We must take care of the land we already have, removing the trees, cultivating the land, and dwelling in it. Only then will God give us more responsibility and more blessing.

Besides the hindrance it is to our future, it is also dangerous to have a standing forest in our lives. Behind the trees, enemies can hide. Then they can turn and conquer us instead of us conquering them. It is vital that we clear out these trees and remove the problems in our own lives.

Tree felling is an honorable profession. The Lord Himself is a tree feller. Let us follow His example. Let us fell the trees that hinder us, and use the cleared land to serve Christ.

50
Tree Worship

The worship of trees may seem antiquated, but it is still practiced in the rural areas of Jordan, Syria, and other Arab countries. Here trees decorated for worship just as described in the Bible can be seen. This worshiping of trees is usually done in honor of the dead, and it is mainly the oak tree that is worshiped.

The prophet Jeremiah commanded Israel not to practice such idolatry in Jeremiah 10. *"Hear ye the word which the LORD speaketh unto you, O house of Israel: Thus saith the LORD, Learn not the way of the heathen, and be not dismayed at the signs of heaven; for the heathen are dismayed at them. For the customs of the people are vain: for one cutteth a tree out of the forest, the work of the hands of the workman, with the axe. They deck it with silver and with gold; they fasten it with nails and with hammers, that it move not. They are upright as the palm tree, but speak not: they must needs be borne, because they cannot go. Be not afraid of them; for they cannot do evil, neither also is it in them to do good." (Jeremiah 10:1-5)*

Some people take this passage as a commandment against having a Christmas tree. They argue that to decorate a tree as described here is to worship it and therefore people should not have Christmas trees. Also the tree in this passage is described as being *"upright as the palm tree."* The palm tree is part of the group of trees known as evergreens, just like the evergreens used for Christmas trees. Because of this verse and its many similarities to the Christmas tree, some sincere Christians do not put up Christmas trees.

But the Christmas tree is part of Christmas tradition, not idol worship. Jeremiah is warning against the worship of idols, not the putting up of a Christmas tree. This is clear from the wording of the passage. He says, *"Be not afraid of them; for they cannot do evil, neither also is it in them to do good." (Jeremiah 10:5)* No one is afraid of the Christmas tree. This is clearly a reference to idol worship.

Again Jeremiah says, *"they must needs be borne, because they cannot go." (Jeremiah 10:5)* This is referring to the way that people carry idols about with them. Still today we see idols carried about, hanging on a car dash or from the rearview mirror. People still carry their idols around with them, whatever they may be. This is a far worse danger than having a Christmas tree.

The Christmas tree is based upon tradition. In reality it, like the holiday of Christmas itself, has nothing to do with Christ. Christ was probably not born on December 25th. It is also probable that the gifts of the wise men were not given on that day, for they probably came when Christ was two years old. Christmas tradition is not based on fact, but on supposition and superstition.

At Christmas time, we often hear the song "O Tannenbaum," which we translate as "O Christmas Tree." The word "Tannenbaum," however, literally means "fir tree" in German. It is translated "Christmas tree" because the fir tree is often used for a Christmas tree.

But many years before the Christmas tree tradition, men worshipped the oak tree. The oak has been held sacred and has been revered by many people throughout history because of its strength, durability, longevity, and hardness. Before the Christmas tree was ever thought of, men used to decorate the oak tree with ornaments as described here in Jeremiah. This idolatrous worship is what Jeremiah is warning against, regardless of the kind of tree.

But it was the Germans who started using the fir tree as a Christmas tree many years later in the 16th century. In fact, this use had a noble origin. Though the Druids worshipped the fir tree, the Germans chose it for an entirely different reason. They chose it because the fir, like most evergreens, is shaped in a triangle. They wanted this triangular shape to teach the concept of the Trinity, with the Father, the Son, and the Holy Spirit at the three points.

In keeping this Christmas tradition, however, the danger is that much of the world may in this way worship the baby of Christmas, but sadly miss the Savior. Luke 2 makes it clear that Jesus Christ is far more than a mere baby in a manger. The angel

judgment, *"O LORD, to thee will I cry: for the fire hath devoured the pastures of the wilderness, and the flame hath burned all the trees of the field." (Joel 1:19)* When fire falls upon the unsaved as in Revelation 8, it is a judgment of total destruction and eternal damnation. But when it falls upon Christians, it is a call to repentance. That is the case here in Joel. Such a fire does not destroy, but rather purifies and helps the trees.

When a fire burns a forested area, it kills the small trees, but not necessarily the large ones. The strong, thick-barked trees can often withstand the fire that consumes the smaller ones. By cleaning out the underbrush and undesirable trees, the fire can actually improve the forested area. In the same way, fiery trials that bring us to repentance improve our lives. They clean out whatever is undesirable and bring us down to what is perfect. Fire purifies gold, and it purifies our lives as well.

A nearby national forest has been suffering from this very problem. It used to be an oak/hickory forest. But it has been so protected from fire that the undesirable trees have grown and begun to take over. It is fast becoming a beech/maple forest. Because it has not been allowed to go through the purifying fire, the desirable trees have diminished, and the undesirable are growing up.

Sadly this seems to be the case with the American church today. Fire throughout history has purified the church. But we have not had to suffer the fire. Thus we have seemingly continued to grow like the national Forest. But it has been the undesirable qualities that have grown. They have crowded out the desirable. We still look like a strong, healthy forest. We have a form of Godliness, but we have no power, no true strength within.

The purifying nature of fire is found throughout the Bible. In Revelation 3, Christ says to the lukewarm church of Laodicea, *"I counsel thee to buy of me gold tried in the fire, that thou mayest be rich." (Revelation 3:18)* Numbers 31 contains the same principle, *"Every thing that may abide the fire, ye shall make it go through the fire, and it shall be clean: nevertheless it shall be purified with the water of separation: and all that abideth not the fire ye shall make go through the water." (Numbers 31:23)*

Not all fire is bad. Instead for us as Christians it is intended for our good, making us produce more fruit. The Jack Pine is a very good illustration of this truth. The pinecone of this tree is sealed very tightly. Even seasonal changes fail to break open the cone and release the seeds. The only force that can break the seal is fire. Only when the Jack Pine goes through a fire are its seeds released so that is may reproduce itself. Only then is it made fruitful. Again we see that fire, which destroys the wicked, calls to repentance and purifies the Christian.

Indeed against His enemies God's fiery judgment is severe. Against them, He uses a "scorched earth policy." Such a policy is used by armies to deny their enemies food resources. The army burns fields and leaves no sustenance for the enemy, even poisoning their water supply. The wicked are God's enemies. For this reason, He will literally scorch the earth in judgment. *"The first angel sounded, and there followed hail and fire mingled with blood, and they were cast upon the earth: and the third part of trees was burnt up, and all green grass was burnt up."* *(Revelation 8:7)* The Lord goes even further and touches the water in verse 11, *"And the name of the star is called Wormwood: and the third part of the waters became wormwood; and many men died of the waters, because they were made bitter." (Revelation 8:11)*

Food and water supplies are now heavily depleted. Still after such severe judgment, these wicked men will not repent. *"And the rest of the men which were not killed by these plagues yet repented not . . ." (Revelation 9:20)* They are beyond repentance. They are the avowed enemies of Jesus Christ.

For these unrepentant sinners, the fire is the fire of destruction. But for the repentant soul, the fire is intended for our purifying and for our good. To such a soul, Christ's offer of Revelation 3 is yet extended, *"I counsel thee to buy of me gold tried in the fire, that thou mayest be rich; and white raiment, that thou mayest be clothed, and that the shame of thy nakedness do not appear; and anoint thine eyes with eyesalve, that thou mayest see." (Revelation 3:18)*

One last judgment on trees we look at is a specific judgment of God upon one man. This judgment is found in Daniel 4.

Mighty king Nebuchadnezzar has dreamed a dream of God's judgment upon a tree, in which the tree is **hewn down**. This hewing down pictures God's judgment upon Nebuchadnezzar for his pride.

Nebuchadnezzar describes what he saw in his dream. *"I saw in the visions of my head upon my bed, and, behold, a watcher and an holy one came down from heaven; He cried aloud, and said thus, Hew down the tree, and cut off his branches, shake off his leaves, and scatter his fruit: let the beasts get away from under it, and the fowls from his branches." (Daniel 4:13-14)*

First the tree is hewn down. But the account does not stop there. The branches are then cut off. Then even the leaves are shaken off the branches. Then the fruit is scattered far from the former tree. And finally the beasts and the fowls get away from it. Every vestige of the former tree is gone, just as every vestige of Nebuchadnezzar's former life would be gone.

One coming upon the site would not even know that the tree had been there, except for one thing. It was commanded that the stump be left in the ground. *"Nevertheless leave the stump of his roots in the earth, even with a band of iron and brass, in the tender grass of the field; and let it be wet with the dew of heaven, and let his portion be with the beasts in the grass of the earth." (Daniel 4:15)*

God reduced proud king Nebuchadnezzar to his lowest estate. God is capable of doing the same to us. If we become lifted up with pride, He will also bring us down to our lowest estate, down to nothing.

In verse 23, Daniel repeats the king's dream to him. *"And whereas the king saw a watcher and an holy one coming down from heaven, and saying, Hew the tree down, and destroy it; yet leave the stump of the roots thereof in the earth, even with a band of iron and brass, in the tender grass of the field; and let it be wet with the dew of heaven, and let his portion be with the beasts of the field, till seven times pass over him." (Daniel 4:23)*

What is this band of iron and brass? Some have speculated that this is referring to a decoration upon the tree that is popular in the Middle East. Often a palm tree or some other tree is decorated with an iron band, like an ankle bracelet on the tree.

But that is not the case here. A tight metal ring on a tree will choke off the nutrients and prevent them from being drawn up the tree, until the ring is removed. This is calling banding, or bonding, a tree. The tree shuts down its nutrients until the band is removed. It is a technique for restricting growth.

Here the Lord is choking off Nebuchadnezzar from returning to reign *"till seven times pass over him." (Daniel 4:23)* This band of iron and brass is a restriction upon him. Such a band will kill a tree if it is on long enough. But in this case, the tree is already gone. The band is placed around the stump to restrict its nutrients and prevent it from growing again until the band is removed.

At times God must bring judgment upon us. Always His desire is that we repent, and His purpose is our good.

52
The Willow Tree

The Latin classification for the willow tree is *Salix*. Unlike some of the trees we have studied like the cedar, the Biblical willow is the same willow with which we are familiar in North America.

There are two Hebrew words translated "willow" in the Bible. The first is *arab* (ערב), possibly the word from which we get our words "Arab" and "Arabic." This word means "to braid," or "to weave," which as we shall see, is one of the uses of the willow. The second Hebrew word is *tsaphtsaphah* (צפצפה), and it is used only once, in the book of Ezekiel. Both words refer to the true willow. However, while some commentaries use the two words interchangeably, there are specific reasons why these two different words are used where they are.

The first mention of the willow tree in the Bible is in Leviticus at the institution of the feast of tabernacles. God commanded, *"And ye shall take you on the first day the boughs of goodly trees, branches of palm trees, and the boughs of thick trees, and willows of the brook; and ye shall rejoice before the LORD your God seven days. . . . Ye shall dwell in booths seven days; all that are Israelites born shall dwell in booths."* *(Leviticus 23:40,42)*

Again we find ourselves looking at these tabernacles, or booths, because their construction involved many trees. These booths were to serve as a reminder to the children of Israel of their origin, that they had begun as mere traveling pilgrims dwelling in temporary booths. These booths were to remind them of God's goodness and deliverance during this vulnerable time of their lives.

For us as Christians, these booths remind us of our pilgrimage here on earth. We are not permanent dwellers upon earth, but only pilgrims passing through to our true home in heaven.

There are many uses for willow wood. It is very pliable and can be bent easily, and so, as its name indicates, it has been woven into baskets, fishing nets, and even wicker chairs. The wood can also be split into small strips and then used like rope to tie items together.

This tying use is probably the use to which the willow was put here in Leviticus in the making of these booths. *"The boughs of thick trees"* were probably used for the main construction of the booths, and then they were probably tied in place with the *"willows of the brook."*

The Hebrew word for bough literally means fruit or offspring. Here it doesn't refer to edible fruit like we think of the term "fruit." Instead it refers to fruit in the sense of offspring, that which the tree produces, its leaves and branches.

The next mention of the willow tree is in Job 40. Here the Lord is telling Job about a beast called *"behemoth."* *(Job 40:15)* He describes his dwelling place by saying, *"He lieth under the shady trees, in the covert of the reed, and fens. The shady trees cover him with their shadow; the willows of the brook compass him about."* *(Job 40:21-22)*

Many commentaries say that this is an elephant or a hippopotamus. But I don't think so due to the fact that verse 17 says, *"He moveth his tail like a cedar."* Neither the elephant nor the hippopotamus have tails like cedars. So this is probably an animal that we in 21st century North America don't know about. The fact that this animal dwelt among the willows reveals that he dwelt in wet areas, for that is where willows grow. So he was probably some kind of water-loving animal.

These willows among which he dwelt were probably either weeping willows or black willows. These are the two most common types of willows, though there are others. Of these two, it is most likely that they were weeping willows, for they would have covered the animal best if he were trying to hide.

The branches of the weeping willow go all the way to the ground and would have made this animal very hard to find. They would have also provided him with shade him all day long so that he would not have had to change places from morning to evening as the sun moved. The branches would have completely enclosed him and kept him shaded and hidden all day.

The next reference to the willow is found in Psalm 137 during the time of the Babylonian captivity. This Psalm records another time of remembrance, just as the feast of tabernacles was a time of remembrance. Only here, the children of Israel remembered, not what God had done **for** them, but what God had done **to** them.

They lamented, *"By the rivers of Babylon, there we sat down, yea, we wept, when we remembered Zion. We hanged our harps upon the willows in the midst thereof. For there they that carried us away captive required of us a song; and they that wasted us required of us mirth, saying, Sing us one of the songs of Zion. How shall we sing the LORD'S song in a strange land?"* (Psalm 137:1-4)

They said they *"remembered Zion."* No doubt they chiefly remembered the focal point of Zion, the temple. Though some of the people were young enough that they had never seen the temple, they had all no doubt heard stories of the glorious temple, and their hearts longed for it once again.

They remembered the days when they had played their harps in the temple. But now that they were in captivity in Babylon, there was no more song in their hearts or lives. So they hung their *"harps upon the willows."*

Willow branches, especially weeping willow branches, are easy to reach, and so they were the natural place to hang the harps. The willows probably also reminded the children of Israel of the feast of tabernacles and of God's past goodness to them. In hanging their harps there, they were probably clinging to a hope that the God Who had brought them through their days in booths in the wilderness would also bring them through these days of captivity and restore them to their homeland.

Was this right for the children of Israel to do, to hang their harps on the willows and refuse to sing? We are told to rejoice always, and so it probably was not exactly right for them to do. God had put them in that trial in order that they might learn by it, and they should have rejoiced in His purpose.

However, before we hasten to condemn them, let us consider what our own response would have been. It is very difficult for anyone to sing a song in the midst of trials. Paul and Silas did sing in prison, but such joy is rare and difficult to maintain. I wonder if any of us would have been able to sing in such a trial as this captivity.

If another nation took over ours and refused to let us worship in our churches, or worse tore them down, or even carried us out of our land, what would we do? Would we be able to sing a song in such circumstances? We must admit it would be very difficult in a time and place like that.

At least these people didn't throw their harps away for good. They hung them on the low branches of the willows where they could be easily retrieved. By doing so, they expressed their faith that God would deliver them and they would sing again. It's a good thing they didn't get rid of these harps, for indeed God's deliverance was coming, and they would need them again.

Let us remember this and never cast our harps away for good, even in the most bitter trial. Let us always maintain a trust that God will deliver us and give us cause for singing again.

Another thing to consider is the reason for these people's weeping. Were they weeping merely over their circumstances, or over the sin that caused those circumstances? We don't know, but possibly it was over the sin, for that was the root problem. We likewise in our trials should mourn over the root problem, our sin, and not merely over our circumstances.

A good sign is that these children of Israel were weeping for Zion, the temple of God, and not for their own houses. To lose one's house is truly devastating, but to lose one's church is more devastating, for it is the more important of the two. Most people don't think that way today. On Sundays, people are often at their houses, raking leaves, working in their yards and gardens, washing their cars, instead of being at church. They place a higher priority on their houses than on the church, the house of God.

It was for this very reason that the Jewish people missed Christ's first coming. They did not have Christ as their first priority, but were busy with their own things. And it is for the same reason that many today will miss Christ's second coming. He is not first place in their hearts and lives. They are not actively watching for Him, but are busy about their own things.

The willow is also associated with weeping in the next passage in which it is found, which records the carrying away of Moab by the Assyrians. This passage says, *"He is gone up to Bajith, and to Dibon, the high places, to weep: Moab shall howl over Nebo, and over Medeba: on all their heads shall be baldness, and every beard cut off. In their streets they shall gird themselves with sackcloth: on the tops of their houses, and in their streets, every one shall howl, weeping abundantly." (Isaiah 15:2-3)*

The willows appear a few verses later, where we are told of these weeping Moabites, *"Therefore the abundance they have gotten, and that which they have laid up, shall they carry away to the brook of the willows." (Isaiah 15:7)*

"The brook of the willows" is literally "the brook of *arab* (ערב)." This could refer to the valley that is called in the Bible "Arabah," for it is from the Hebrew word *arab* that we get the word "Arabah." Arabah is a low plain south of the Dead Sea, between the Dead Sea and the Red Sea.

Or this brook could be the same river where the Israelites wept and hung their harps, which was probably the river Chebar in northern Babylon. Chebar is about the only low, wet area in Babylon, which is mostly desert.

Wherever this place was, we do know that it was a low plain with a brook running through it. It was to this place that the Moabites' substance was brought.

Some say that Moabites' captors, the Assyrians, carried this substance here, taking it away from the Moabites and storing it here. But the verse says that they themselves, the Moabites, carried their own substance here. Rather than this being a place of storage by the captors, it was probably a place of hiding from the captors. Willows are ideal for this, as we saw with behemoth hiding in the enclosure of the willows.

They might have also brought their substance here because of the transportation that is available on a waterway. Perhaps they hoped that if a ship were to come by, they could put their goods on it and send their substance to some safe place outside of the land of their captivity. In this way, the Moabites perhaps hoped to keep their abundance safe.

As Christians, with our true home and treasure in heaven, we never need to hide our earthly treasures in this way, for we know that they are just that, earthly. We should not be foolish or unwise in the way we handle our earthly goods, but neither should we hoard them or be overly protective of them, keeping them in a safe and constantly fearing thieves and robbers.

We should realize that all we own is really the Lord's anyway, that He knows full well how to protect His goods, and that He can repay tenfold anything that is lost. In fact, rather than hiding our money from others, we should spend it on others and give it to them.

The next reference to the willow tree is also in Isaiah. Here the Lord pronounces a blessing upon Israel, saying, *"For I will pour water upon him that is thirsty, and floods upon the dry ground: I will pour my spirit upon thy seed, and my blessing upon thine offspring: And they shall spring up as among the grass, as willows by the water courses."* (Isaiah 44:3-4)

Just as literal water is poured on a literal seed, so God promised to pour His Spirit on the seed of Jacob. In John 7, Jesus compared the Holy Spirit to water when He said that the Spirit would flow from the believer like *"rivers of living water."* (John 7:38) When this water of the Spirit would come upon the seed of Jacob, they would spring up *"as willows by the water courses."*

Willows are very fast-growing trees, growing four to five feet per year. This means they are not long-lived, but they do illustrate well the principle of growth which is our focus here. Just as they grow fast when supplied with abundant water, so Israel and we as Christians grow fast when supplied by God's Spirit. God's Spirit provides all that is necessary for our growth.

It is interesting here that Israel is likened to a willow tree. Willow wood, like Israel, is very diverse in its uses. Willow wood is used to make something as trivial as a cricket bat, which is popular in England. But it has also been used for something as serious as a soldier's shield and gunpowder.

The Romans used to make their shields out of willow wood because of how easily it could be bent, formed, and molded. For gunpowder, willow wood is charred into charcoal, which is then made into the gunpowder. Thus willow wood is a very versatile substance, and again we see that the willow is a tree whose wood is more important than its fruit.

With its versatility, willow wood is a picture of the nation of Israel. Willow wood can be used both for joyful play (cricket) and for war. And in the same way, Israel can be both a source of joyful blessing and a curse. Zechariah presents this paradox well, saying, *"And it shall come to pass, that as ye were a curse among the heathen, O house of Judah, and house of Israel; so will I save you, and ye shall be a blessing: fear not, but let your hands be strong." (Zechariah 8:13)*

The Gospel of Jesus Christ also has this twofold function. It can be a great encouragement, but it can also be a two-edged sword, piercing and cutting to convict of sin. Just like the willow, it can be a source of great delight and also a source of great pain over sin.

The last mention of the willow in the Bible is in Ezekiel 17, the passage we looked at in detail in chapter 16. We will now focus just on the willow in this passage.

In this passage, the Lord puts forth a parable. *"Thus saith the Lord GOD; A great eagle with great wings, longwinged, full of feathers, which had divers colours, came unto Lebanon, and took the highest branch of the cedar: He cropped off the top of his young twigs, and carried it into a land of traffick; he set it in a city of merchants. He took also of the seed of the land, and planted it in a fruitful field; he placed it by great waters, and set it as a willow tree." (Ezekiel 17:3-5)*

This willow tree is set in an ideal location. It is planted *"in a fruitful field"* where it has all the nurturing it needs, and it is placed *"by great waters"* where it has all the water it needs. This matter of water is especially important to the water-loving willow tree.

This is the only place where the second word for the willow is used, *tsaphtsaphah* (צַפְצָפָה). This word comes from a root that means "to overflow" and is applied to this water-loving tree because it grows in the floodplain where the waters overflow. This unique word is used here because of the unique features of this passage. This is the only place in the Bible that specifically refers to a singular willow tree. All the other passages simply speak of willows. Also this passage is a parable or a riddle.

The eagle in this parable is Nebuchadnezzar. He carried away the cedar, which was Jechoniah the king of Judah. He then took *"the seed of the land,"* which refers to the royal seed, Zedekiah the brother of Jechoniah. And he made this seed, Zedekiah, the new king of Judah in the place of Jechoniah.

But he didn't make this new king strong and permanent like the cedar had been. Instead he made him like a willow tree. A willow tree doesn't live nearly as long as a cedar, it is not as strong as the cedar, and it does not get as tall as the cedar. Nebuchadnezzar didn't want Zedekiah to be king for long. He wanted him to be only a temporary and weak king, under full subjection to Babylon.

As it turned out, Zedekiah realized his weak position and tried to get out of it by making a league with Egypt, when he was supposed to be in league with Babylon. But Babylon then found out about this secret league and came to destroy him. Under this pressure, when it really came down to it, Egypt refused to fulfill her end of the league.

Now Zedekiah was forsaken by Egypt and out of favor with Babylon. Thus by his own choice, Zedekiah caused the water of the Holy Spirit to dry up from him, and he withered like a willow tree when it is deprived of water.

53
Wormwood

Wormwood is not a substance with which most of us are very familiar. In fact, many people think it is only a figurative term and not a real substance. Or when they find that there is a real plant named wormwood, some still think that the term "wormwood" was originally figurative in the Bible and that the plant got its name from this Biblical figure. But in reality, wormwood has been the name of a plant for many years. Egypt used the term "wormwood" for this plant in the first century B. C. So it has been around, but it is just not common and not normally used.

Wormwood is mentioned in both the Old and the New Testament. Its Hebrew name is *la'anah* (לענה), and its Greek name is *apsinthos* (αψινθος). Throughout the Bible, it is always mentioned in a bad light, like another plant we have looked at, hemlock. Wormwood is always poisonous and bitter. In its pure form, it is extremely poisonous and will definitely kill a person. For this reason, it symbolizes bitterness and cursing and death.

Wormwood is not a true tree, but an herb. But since it contains the term "wood," we will include it in our study. We did the same with gopher wood. Though it is not a kind of tree, we included it because it contains the term "wood." Wormwood, far from being a tree, is really in the daisy family. Its scientific Latin name is *Artemisia absinthium*, the latter word being a derivative of the Greek word *aspinthos* (αψινθος).

The first Biblical mention of wormwood is in Deuteronomy where Moses is reminding the people of their journey through the pagan nations that serve false gods. He is warning them not to do the same, and says, *"(For ye know how we have dwelt in the land of Egypt; and how we came through the nations which ye passed by; And ye have seen their abominations, and their idols, wood and stone, silver and gold, which were among them:) Lest there should be among you man, or woman, or family, or tribe, whose heart turneth away this day from the LORD our God, to go and serve the gods of these nations; lest there should be among you a root that beareth gall and wormwood." (Deuteronomy 29:16-18)*

Here we find wormwood mentioned with gall, as it almost always is in the Bible. Since this is the case, we will examine gall as well in this chapter, and compare and contrast it with wormwood. Gall is always bitter, but its identity is a little unclear. It could be a plant itself, it could be berries, or it could even be venom from a reptile. But personally I don't believe it's poisonous, but only bitter.

I believe that gall is made of myrrh because of the descriptions of the drink offered to Jesus on the cross. Matthew says that He was given vinegar mingled with gall, but Mark says that He was given myrrh mingled with wine. So in order for both to be true, I believe that gall is a mixture of myrrh and other liquids.

Myrrh itself, like the gall it makes, is always bitter. Its Hebrew word is *more* (מֹר), which means "bitterness." It is used in coffins and to anoint dead bodies.

But still, myrrh is not poisonous. If it were poisonous, then historians and skeptics today would say that Christ died of being poisoned on the cross, rather than the truth that He gave His Own life willingly. Myrrh, though it is bitter, is not poisonous.

This brings up the question of why the wise men brought myrrh to the child Jesus. Why would they bring a tiny child something bitter? Perhaps the wise men themselves did not even realize it, but this myrrh was a symbol of the bitterness that Jesus would one day suffer for our sins on the cross. Indeed in that day, His body would be anointed with myrrh when it was taken down from the cross and laid in the tomb. Thus myrrh appeared at both His birth and His death.

This allows us to characterize the three unpleasant Biblical plants often grouped with one another, gall, wormwood, and hemlock. Gall is bitter, but not poisonous. Hemlock is poisonous, but not bitter. And wormwood is both bitter and poisonous. So of the three, wormwood is the worst. These three are all similar in their Biblical occurrences. They are never good, but are always symbols of bad things, such as bitterness and cursing.

To get the full sense of the symbolism of these substances in their contexts, it is sometimes helpful to read the verses by substituting the names of these substances with their meanings. For example, the verse we have read in Deuteronomy would read, "lest there should be among you a root that beareth bitterness, and poison and bitterness." Thus this passage is warning that turning away from the Lord always brings bitterness.

The next mention of wormwood is in Proverbs 5. In this passage Solomon is speaking of the strange woman, and he is probably speaking from his own experience, for the Bible says that he *"loved many strange women." (I Kings 11:1)* Solomon warns, *"For the lips of a strange woman drop as an honeycomb, and her mouth is smoother than oil: But her end is bitter as wormwood, sharp as a twoedged sword." (Proverbs 5:3-4)*

Solomon here presents a contrast between the sweet honeycomb and the bitter wormwood. He says that the strange woman begins sweet like the honeycomb, but she ends *"bitter as wormwood."*

Sin and Satan are also like that. They at first send out sweetness and make themselves look good, but their end is bitter. No matter how sweet and attractive Satan makes sin look, he always gives bitter fruit at the end. The sweetness won't last.

But Christ is exactly the opposite. He sends bitter trials and problems to His Own children, but these very trials turn to sweetness in the end. For us who trust Christ, we have the ultimate sweetness awaiting us, a home in heaven that will make the bitterest trial here on earth look like only a drop in the bucket.

Let us take this warning from wise Solomon. Sin may look enticing, but in the end it is not at all. In the end, we will wish we had never gotten into it, never even touched it. Like wormwood, it will end in bitterness and death. For wormwood is always a symbol of bitterness, cursing, and death, never life.

The book of Jeremiah is the next place where wormwood is mentioned, and it mentions it twice. The first time, it is speaking of the children of Israel in general. The Lord says that they *"have walked after the imagination of their own heart, and after Baalim, which their fathers taught them." (Jeremiah 9:14)* For this sin, God pronounces judgment. *"Therefore thus saith the*

LORD of hosts, the God of Israel; Behold, I will feed them, even this people, with wormwood, and give them water of gall to drink." (Jeremiah 9:15)

Again we see wormwood and gall together. This speaks of both bitterness and death. Indeed because they have followed Baalim and the imaginations of their own hearts, God will feed His people with the bitterness of affliction and very grievous tortures. The scattering and persecutions and horrible deaths suffered by the Jewish people throughout history have been a fulfillment of this verse, the judgment of God upon their sin.

The second mention of wormwood in Jeremiah is similar to this one, but a little different. This time the focus is not on the general populace, but on the prophets in particular. The Lord says, *"I have seen also in the prophets of Jerusalem an horrible thing: they commit adultery, and walk in lies: they strengthen also the hands of evildoers, that none doth return from his wickedness: they are all of them unto me as Sodom, and the inhabitants thereof as Gomorrah. Therefore thus saith the LORD of hosts concerning the prophets; Behold, I will feed them with wormwood, and make them drink the water of gall: for from the prophets of Jerusalem is profaneness gone forth into all the land." (Jeremiah 23:14-15)*

These prophets have prophesied falsity and thereby made the people to sin. But not only have they made the people to sin, but they have sinned themselves. They have not only taught lies, but believed in the lies themselves. In other words, they have spoken bitterness and death through their mouths to the people, and practiced bitterness and death in their own lives. This has become bitterness in the mouth of God, and therefore God is now going to feed them with their just reward, bitterness and death.

Jeremiah, called the weeping prophet, mentions wormwood twice again in his other book, Lamentations. This time the two are just a few verses apart and in the same context. Jeremiah laments over sin, saying, *"He hath filled me with bitterness, he hath made me drunken with wormwood. He hath also broken my teeth with gravel stones, he hath covered me with ashes. And thou hast removed my soul far off from peace: I forgat prosperity. And I said, My strength and my hope is perished from the LORD:*

Remembering mine affliction and my misery, the wormwood and the gall." (Lamentations 3:15-19)

Jeremiah describes this man as *"drunken with wormwood."* He is drunken with bitterness and death, and so, like all drunken men, he doesn't know which way to turn. Sin makes the cup of affliction very bitter, and sin always brings death. In general terms, sin brought Adam's eventual death, and sin has brought death to all men. But in more specific terms, sin also brings individual death. Sin in our lives always brings bitterness and death, not only to ourselves, but to all around us.

In the last verse of this passage, this man remembers *"the wormwood and the gall."* He remembers the bitterness and death of sin. This man has found what Solomon found, that sin is sweet at first, but very bitter in the end. This man now remembers the great bitterness that sin has brought to him.

We all do this with food. Once we taste a food that is bitter, we never forget that taste. We remember how bitter it was and want to stay away from it.

But sometimes, even though we know how bitter it is, we go back to it. I am that way with real spicy food. After a while of not eating it, I don't remember how hot it was, and I go back to it. Of course, that reminds me of how hot it is, and I stay away from it again. I do the same thing with White Castle hamburgers. After not having them for a while, I go back and eat them and am reminded all over again why I don't like them.

Tragically we often do the same with sin. After experiencing the bitterness of sin, we determine never to repeat it. But after a time, we forget how bad it really was, and we go back to it. Then we experience its bitterness again and say to ourselves, "Oh, that's why. I won't go back to it." But again, after a time, we say, "That wasn't that bad," and we repeat it again. We just don't learn.

The book of Amos gives us the next Biblical mention of wormwood. Amos said, *"For thus saith the LORD unto the house of Israel, Seek ye me, and ye shall live: But seek not Bethel, nor enter into Gilgal, and pass not to Beersheba: for Gilgal shall surely go into captivity, and Bethel shall come to nought. Seek the LORD, and ye shall live; lest he break out like fire in the house of Joseph, and devour it, and there be none to quench it in Bethel. Ye who turn judgment to wormwood, and leave off righteousness in the earth." (Amos 5:4-7)*

Amos mentioned Bethel and Gilgal, recalling the past history of Israel. This would have brought to the minds of the people the days of Moses and Joshua, and would have perhaps reminded them of Moses' solemn warning not to turn away from the Lord, lest there should be among them *"a root that beareth gall and wormwood." (Deuteronomy 29:18)*

Yet Amos was here to tell them that the warning of Moses had gone unheeded, and that the root had now borne gall and wormwood. The children of Israel in this day were doing what is known as *corruptio optimi est pessima*. They were doing wrong under the pretence of doing right. They seemed to be practicing judgment, but it was only a pretence. They instead turned what should have been right judgment into a bitter curse, which in turn brought down God's bitter curse upon them.

This verse is very similar to another verse in the next chapter of Amos, which says, *"Ye have turned judgment into gall, and the fruit of righteousness into hemlock." (Amos 6:12)* In the first verse, judgment was turned into wormwood; in the second, it was turned into gall. In the first verse, righteousness was left off in the earth; in the second, it was turned into hemlock.

This verse also, like the first, speaks of Israel practicing *corruptio optimi est pessima*, doing wrong in the disguise of doing right. So the children of Israel had doubly committed this sin. They had turned judgment into both wormwood and gall, and they had both left off righteousness and turned it into hemlock. So just as they had doubly committed their sin, so God would doubly judge them.

The children of Israel had been saying that they were doing judgment, righteousness, when all the time they had known they were doing wrong. They had left off righteousness, causing it to take a back seat. Therefore, they may have had a form of judgment, but they had no justice. Justice and righteousness had nothing to do with their judgment. This turned even their seemingly good practice, judgment, into the bitterness and death of wormwood.

The last mention of wormwood is in the last book of the Bible. It is at the sounding of the third trumpet in Revelation. *"And the third angel sounded, and there fell a great star from heaven, burning as it were a lamp, and it fell upon the third part of the rivers, and upon the fountains of waters; And the name of the star is called Wormwood: and the third part of the waters became wormwood; and many men died of the waters, because they were made bitter." (Revelation 8:10-11)*

The first mention of wormwood in this passage is as the name of a star, not a plant. This star is called Wormwood because it is like wormwood.

There is some debate over what this star is. Stars in the Bible are sometimes pictures of angels, and so this could be an angel, perhaps an angel of Satan or some other angel. We are told that Satan swept the third part of the stars, presumably the third part of the angels. But though stars are angels a lot of times in the Bible, they are not always.

Some commentaries offer another possibility, that this star is a political or religious person, because of the phrase *"burning as it were a lamp."* They also say this because a literal star striking the earth would fragment and destroy the earth, therefore it must be a figurative term for a person. A star is much bigger than a comet, and it is easy to understand this reasoning. But still, we know that the omnipotent God could make even a literal star hit the earth and not affect it. He could make it fragment when it hit the atmosphere before it ever hit the earth, for God can overcome any natural or physical forces and can do anything He wants to.

So exactly what this star is, we don't know. We do know that it is called *"a great star,"* so it is not just any star. We also know that it is as wormwood, as bitterness and death which is seen in its effect. *"The third part of the waters became wormwood; and many men died of the waters, because they were made bitter." (Revelation 8:11)*

This is similar to what happens after the sounding of the second trumpet just a few verses earlier. *"And the second angel sounded, and as it were a great mountain burning with fire was cast into the sea: and the third part of the sea became blood." (Revelation 8:8)* At this second trumpet, a third part of the sea becomes blood due to the burning mountain, whereas at the third trumpet, a third part of the rivers become wormwood due to the burning star.

This leaves not much water that the plants of the earth can use. Indeed the Tribulation is a time that brings great harm to the vegetation of the earth. It is not a time that is advantageous for planting a garden or growing plants. From other passages, we learn that the earth is burned and scorched as well, further hurting the plants.

But not only is the Tribulation harmful to vegetation, but also to the people of the earth. They are here left with very little water that they can drink. They have the choice of drinking either blood or wormwood. In spite of the fact that wormwood is both bitter and poisonous, the people probably drink it because it looks better than blood. This leads to certain death, and we are told, *"Many men died of the waters, because they were made bitter."* *(Revelation 8:11)*

Here we find bitterness and death together once again, for that is always the twofold effect of wormwood. It is both bitter and poisonous. In this way, it teaches us the most solemn lesson of the three substances we have examined, hemlock, gall, and wormwood.

Hemlock, which is poisonous but not bitter, teaches us the lesson of Solomon. Some things may taste sweet, but they bring death in the end.

Gall, which is bitter but not poisonous, teaches us a vital lesson for the Christian life. There are some things that will come our way that are bitter, but won't kill us. We see this in other foods as well. For example, the young leaves of watercress have a bitter taste, but they are certainly not poisonous. Also, a little known orange is the Seville orange, which is very bitter. Because of its bitterness it is unpopular, but in spite of its bitter taste, it is not poisonous.

But wormwood is both bitter and poisonous. It brings certain death if you eat it. It teaches us the very sobering, but also very important, lesson that sin brings only bitterness and death.

On this solemn note is an appropriate place to end our study. We have ranged throughout the Bible, from the first book,

Genesis, to the last book, Revelation, and we have covered the trees of the Bible from the tree of life to the plant of death, wormwood. Let us learn its lesson well. Let us turn from our sin, which brings only bitterness and death, and let us turn to Jesus Christ, the true tree of life, Who gives eternal life to those who trust in Him.

Bibliography

Clarke, Adam. *Adam Clarke's Bible Commentary.*

Allen, Ward. *Translating for King James.* Vanderbilt University Press, 1993.

Al-Mawrid. "Actual Site of Baca/Bakkah." *Foundation for Islamic Research and Education.*

Balfour, John Hutton. *The Plants of the Bible, Trees and Shrubs. Books.google.com.*

Bible Desk. P.O. Box 10142. St. Petersburg, FL 33733.

Bible History Online. 2229 NE Burnside Street, Suite 43. Gresham, Oregon 97030. 2012. *www.bible-history.com.*

Bodains, David. *The Secret Garden: Talking Beetles and Signaling Trees: The Hidden Ways Gardens Communicate.*

Chancey, William. "Cedar of Lebanon." *Arbor Age Magazine* 1993.

Chen, Rujin. Silver, David L. and De Bruijn, Frans J. *Plant Cell.* 1998. 10:1585-1602. doi:10l1105. Tpc. 10.10.1585.

Christie, W. M. Church of Scotland. *The Barren Fig Tree.* Palestine Under British.

Cloud, David. *The Glorious History of the King James Bible.* Way of Life Publishing, 2008.

Easten Bible Commentary.

"81 Useful Interesting Facts About Trees." *American Wood Source*. 2011.

"Fig Tree." *International Standard Bible Encyclopedia*. Volume II. Page 302.

Fragman-Sapir, Dr. Ori. Jerusalem Botanical Gardens. *The Tree That Nearly Disappeared From the Country*. 2012. Tal1000@gmail.com.

Hardy, Barry. *Chemin Fostreom*. September 24, 2011.

Harlow, Harrar, and White. *Text Book of Dendrology*. 9th Edition. 2000.

Haughwont, Mark. *Chronicles and Kings, A Comparison*. Jerusalem: Hebrew University, 2002.

Henry, Matthew. *Matthew Henry's Commentary*. Zondervan Publishing, 1961.

Hinton, John, Ph.D. *Pitch of the Ark*. Bible Restoration Ministry. AV1611.com/KJPp.

Johnson, Kristine. *Populus Alba.* Plant Conservation Alliance. Alien Plant Group. 07-01-2009.

Kaiser, Davids, and Bruce. *Hard Sayings of the Bible*. 1996.

Knee, Michael. "Plant Cells Article." *Ohio State University, Horticulture and Crop Science, Division*.

MacPherson, J. M.A. Menzies, J. *Biblios Lesson*.

Madison, Deborah. *Edible, World Food Plants*. National Geographic, 2008.

Mangum, Douglas. *The Bible Hebraica*. 2008-2012.

Mazzaro, Maria. "Best of Sicily." Magazine 2005.

McDermott, Jeanne. "The Golden Spruce." *Smithsonian Magazine*. Vaillant, John. December 1984.

Mendez, Sr., Arnold C. 222 Pearse Drive. Corpus Christi, TX.

Moldenke, H. N. & A. L. "Plants of the Bible." Article 175. Reama (Forsk) Web and Berth.

Morris, Henry, Ph.D. *Tree of Life*. Institute for Christian Research.

Musselman, L. J. *Bible Plants*. Norfolk: Old Dominion University, 2008.

Musselman, L. J. "Trees in the Koran and the Bible." *Unasylva* 213. Volume 54. 2003.

Norton, David. *A Textual History of the King James Bible*. Cambridge University Press, 2005.

Perez, Alano. "Allah the Oak?" *Answering Islam*. 1999-2012.

Research Gate Corporation. 285 Third Street #727. Cambridge, MA 02142.

Rogers, Adrian. "Don't Settle for less Than God's Best." *Love Worth Finding*. 2457CO.

Russell, Cutler, and Walters. *Trees of the World*. Anness Publication Ltd., 2006, 2007.

Schultz and Baldsin. "Hey, Poplars Talk." *Science Magazine* 1979.

Schurrie, H. "The Fig." *Timber Press Horticulture Review* 12:499 (1990).

Shmida, Avi. *Guide Book of Trees of Israel.*

Spurgeon, C. H. *Withered Fig Tree.* Sermon #2107. September 29, 1889.

Sternberg, Guy. "What Is This Thing Called Oak?" *ISA Bulletin* 1998.

Strong, James. *Strong's Exhaustive Concordance.* Old Time Gospel Hour Publishing.

Talmage, Dr. T. DeWitt. *Religion in Booths.*

Taylor, Paul S. *Christian Answers Network.* P.O. Box 200. Gilbert, AZ 85299. 2001.

Abernethy. Turner, R.E. *U.S. Forest Wetlands 1940-1980.* Bio-Science 37 (10). 721-727. 1987.

Unger, Merrill. *Unger's Bible Dictionary.* Moody Press, 1980.

Worchester, John. *Plants.* Science of Correspondences, 1875.

Index

About the Author

William Mitchell's professional record begins with AAS, BS, and MS degrees in Forestry and Environment Science from Southeastern Illinois College and Pacific Western University from 1973-1980. He served as an instructor at Southwestern Illinois College from 1993 to 1998 in the horticulture department. For thirty-five years he served as a forestry consultant for several cities in the Midwest. He served as the president of the Regional Urban Forestry Council from 1998 to 2008. He has also served in various churches over these same fifty years as a church worker and leader, and has served as a deacon for twelve years.

Mitchell's Bibles are full of marginal notes that point out scientific and practical uses of the many Bible passages that mention trees. It is his life-long desire to combine dendrology and theology, studies that complement rather than conflict with one another. So it is to you who now read this book that his highest professional purpose is dedicated, that he might help each of us live fruitful, strong, and vibrant spiritual lives for Jesus Christ.

CPSIA information can be obtained at www.ICGtesting.com
Printed in the USA
LVOW120132070513

332411LV00001B/5/P

9 781937 129699